Public Health
Entomology

Public Health Entomology

2nd edition

Jerome Goddard

CRC Press
Taylor & Francis Group
Boca Raton London

CRC Press is an imprint of the
Taylor & Francis Group, an **informa** business

Cover design: Laura C. McKee

Photos and specimens: U.S. Air Force C-130 spraying for mosquitoes after Hurricane Katrina. (USAF photo courtesy of Paul Schwarzburg and Lt. Col. Mark Breidenbaugh.) Flea specimen. (Courtesy of Lawrence Bircham.) Mosquito feeding. (Photo courtesy of the Centers for Disease Control.) Larval tick. (Photo by Jerome Goddard.) Hard tick. (Photo courtesy of Dr. Lorenza Beati, U.S. National Tick Collection, Statesboro, GA.)

First edition published 2022
by CRC Press
6000 Broken Sound Parkway NW, Suite 300, Boca Raton, FL 33487–2742

and by CRC Press
2 Park Square, Milton Park, Abingdon, Oxon, OX14 4RN

CRC Press is an imprint of Taylor & Francis Group, LLC

© 2022 Jerome Goddard

Library of Congress Cataloging-in-Publication Data
Names: Goddard, Jerome, author.
Title: Public health entomology / Jerome Goddard.
Description: Second edition. | Boca Raton, FL : CRC Press, 2022. | Includes bibliographical references and index. | Summary: "This second edition of Public Health Entomology maps onto Certificate courses in public health entomology offered by universities and U.S. Centers of Excellence. It comprehensively examines vector-borne disease prevention, surveillance, and control from a governmental and public health perspective with worldwide application"— Provided by publisher.
Identifiers: LCCN 2021043516 (print) | LCCN 2021043517 (ebook) | ISBN 9780367636470 (hardback) | ISBN 9780367636463 (paperback) | ISBN 9781003120087 (ebook)
Subjects: LCSH: Insects as carriers of disease. | Entomology. | MESH: Insect Control—methods | Insect Vectors—pathogenicity | Insect Bites and Stings—complications
Classification: LCC RA639.5 (print) | LCC RA639.5 (ebook) | NLM WA 110 | DDC 616.9/68—dc23
LC record available at https://lccn.loc.gov/2021043516
LC ebook record available at https://lccn.loc.gov/2021043517

ISBN: 978-0-367-63647-0 (hbk)
ISBN: 978-0-367-63646-3 (pbk)
ISBN: 978-1-003-12008-7 (ebk)

Typeset in Palatino
by Apex CoVantage, LLC

Dedication

*To my graduate students **(Minor Professor):**
Flavia Ferrari (Ph.D.), Jessica Aycock (M.S.),
and Jillian Masters (M.S.). **(Major Professor):**
Kristine Edwards (Ph.D.), Lauren Goltz (M.S.),
Gail Moraru (Ph.D.), Wendy Varnado (Ph.D.),
Santos Portugal (Ph.D.), Tina Nations (Ph.D.),
Sarah McInnis (M.S.), Grant de Jong (Ph.D.),
Michelle Allerdice (Ph.D.), Alyssa Snellgrove (Ph.D.),
and Afsoon Sabet (M.S.).
More than you know, all of you have enriched my life
and Rosella and I cherish the times we had with you.*

Contents

Foreword

Finally, there is a reference text on public health entomology. There have been numerous books and (now) online resources about the biology, ecology, and control of medically important arthropods, but nothing that comprehensively covers public health entomology (as opposed to academic medical entomology), vector-borne disease prevention, surveillance, and control. During my career as a state epidemiologist and eventually as state health officer, I saw firsthand the devastating effects of arthropods and the diseases they can cause. Throughout the years, we investigated and tried to manage disease clusters and outbreaks due to Rocky Mountain spotted fever, St. Louis encephalitis, eastern equine encephalitis, and West Nile virus. I personally have visited with grieving family members of those who died from these diseases as well as patients who survived but now are permanently disabled, so I don't need to be convinced that entomology is an important piece of the puzzle of public health and effective disease prevention. Some of the earliest public health efforts to better health and well-being were geared toward filth fly and mosquito control, and often included basic sanitation measures, such as installation of screen-wire windows and doors, clean water supply, sewage disposal, and the elimination of insect breeding sites. Modern public health still employs these tools and methods, as well as the addition of immunizations, (better) pesticides, and high-tech disease surveillance and control techniques.

Jerome Goddard's experience and talent make him perfectly suited to write a book about public health entomology. For many years, I worked with Dr. Goddard when he was our public health entomologist and found him to be extremely knowledgeable in his understanding of both the bugs and their habits (we called him the "Bug Man"), as well as how they fit into the environment, and the life cycles of vector-borne disease. In addition, he designed and implemented the vector control program along the Mississippi Gulf Coast after Hurricane Katrina, adding disaster vector control to his experience and list of accomplishments. His wonderful ability to communicate that knowledge and experience, both verbally and through the written word, to the rest of us in public health and to the general public has prevented an unknowable amount of disease in Mississippi.

This book is divided into two main sections: (I) the basic principles of assembling and operating a public health entomology program, and (II) a brief overview of the main public health pests. The first section is particularly useful in that it provides practical advice on how to set up a public health entomology program from scratch and includes chapters on operational research, funding opportunities, and (to my delight) where to go for help. To my knowledge, there is no such information available anywhere. In addition, there are chapters on the history of public health and entomology, disaster vector control, and vector-borne disease surveillance. One of the more interesting chapters, "Regulatory and Political Challenges," is sure to evoke much thought and discussion among health department personnel nationwide, as they regularly deal with such issues.

Infectious diseases, including those that are vector-borne, are here to stay. Although not vector-borne, the covid-19 pandemic was a wake-up call for public health officials. International travel, climate change, military activity in remote and tropical parts of the world, and even the spreading of suburbs will likely lead to an increase in vector-borne disease incidence and distribution in temperate zones. Public health officials must be ready to meet these new challenges. This text should go a long way in helping health officials to prepare for and respond to these future threats.

Mary Currier, MD, MPH
Former State Health Officer
Mississippi Department of Health
Jackson, Mississippi

Preface

Public health disease control efforts are immensely valuable. Historically, infectious diseases exacted a heavy toll on human societies. For example, in a small family cemetery plot near Natchez, Mississippi, the tombstones reveal that 7/8 (88%) of one couple's children died by the age of 18. These deaths were likely a result of childhood infectious diseases, based on the seasonal patterns of the dates of death (on separate occasions two or more children died within days of each other). We might speculate that these deaths were due to yellow fever as it occurred in several waves up and down the Mississippi River during the 1800s, but nonetheless, this family was devastated by infectious diseases.

Despite what some people think, infectious diseases, and specifically (the subject of this book), vector-borne diseases, have by no means been conquered in modern times. In fact, good arguments can be made that they are winning the war for survival against humans. Diseases continue to emerge and reemerge worldwide, and many of them are vector-borne. Take Chikungunya (CHIK), for example. This disease, which originated in Africa, is carried by mosquitoes and causes intense pain in the joints that can last for months or years. Results of recent research show that CHIK's rapid spread in Asia was launched by a single mutation in an African strain of the virus. The alteration was seemingly insignificant—a single amino acid change in one of the virus's exterior envelope proteins—leading one researcher to compare it to "a single missing comma in a six-page short story." But this mutation enabled the virus to efficiently infect *Aedes albopictus*, a species of mosquito found nearly worldwide (and common in the United States). The mutated CHIK virus took full advantage of its new host, infecting millions of people as it spread across India, Thailand, and Malaysia. There are many other examples of emerging or reemerging vector-borne diseases, such as dengue fever, which continues to ravage much of Central and South America and the Caribbean. Of special concern to people living in the United States is the establishment of locally acquired dengue cases in 2009 and 2010 in Florida. Further spread of dengue in the southeastern United States is likely.

In the struggle against vector-borne diseases there is a great need for information that can bridge the gap between vector control workers on the ground (practitioners) and public health program planners and administrators. The Centers for Disease Control and the World Health Organization have provided limited guidance in this niche, which is freely available online; however, no comprehensive reference books are available. Although there are a number of very good texts available on the biology, ecology, and control of vector-borne diseases, there is no book that comprehensively looks at vector-borne disease prevention, surveillance, and control from a governmental or public health perspective. That is precisely what *Public Health Entomology* aims to do, and I hope you find it useful. As always, I solicit your comments and feedback. They will help me refine the work for future editions.

Jerome Goddard, PhD
Mississippi State University
Starkville, Mississippi
jgoddard@entomology.msstate.edu

Acknowledgments

First Edition

Jeanette C. Martinez, U.S. Environmental Protection Agency, Washington, DC, provided information on the pesticide registration process from the Office of Pesticide Programs (OPP), and Flavia Girao, a PhD student at Mississippi State University, provided the overview of WHO vector programs. Much of the Mississippi Malaria section in Chapter 1 comes from a paper coauthored with Dr. Kelly Hattaway and published in the *Journal of the Mississippi Academy of Sciences* and is used with permission. Maj. Mark Breidenbaugh, U.S. Air Force, provided information about the military aerial spraying capability and response after Hurricane Katrina. Dr. Brigid N. Elchos, Deputy State Veterinarian (Mississippi), and Dr. Abelardo Moncayo, Tennessee Department of Health, kindly provided much needed comments and suggestions on several earlier versions of this book. Parts of Chapter 4 concerning mosquito surveillance and habitat monitoring came from a health department publication developed by Mr. Ed Bowles. Many of the photographs used in the book are from Dr. Blake Layton (Mississippi State University), Wendy C. Varnado (Mississippi Department of Health), and Stoy Hedges (Terminix, Inc.). I especially thank Drs. Mary Currier, Sally Slavinski, Alan Pennman, Tom Brooks, and all my other friends, either currently or formerly employed at the Mississippi Department of Health. Everything I know about public health I learned from them.

Second Edition

Michelle Allerdice (Centers for Disease Control, Atlanta, GA) provided information for the Inception of the Centers for Disease Control section in Chapter 1, and Sarah McInnis provided the original CDC organization chart figure. Terry L. Carpenter (Armed Forces Pest Management Board) kindly supplied information about the history of military medical entomology. The section on State Public Health Veterinarians in Chapter 3 is taken in part from an article on the website of the National Association of State Public Health Veterinarians (Nasphv.org) originally written by a friend of mine, the late Dr. Bill Johnston. The section on genetically modified mosquitoes was originally written by my M.S. graduate student, Afsoon Sabet. Gene Merkyl and Lois Connington (Mississippi State University Extension Service) provided much-needed information on pesticide certification and licensing. The current status of National Pollutant Discharge Elimination System (NPDES) permits and mosquito control was accessed from the American Mosquito Control Association website. Tammy Jackson (Public Works Department, City of Vicksburg, MS) allowed me to photograph some of their mosquito control equipment. Stoy Hedges (Stoy Hedges Consulting Company) graciously allowed me to use some of his wonderful photographs, as did Dr. Blake Layton (Mississippi State University) and Dr. Wendy Varnado (Mississippi Department of Health). My wife, Rosella Goddard, digitally created worldwide distribution maps for the various arthropod diseases, and I am especially grateful for that.

About the Author

Jerome Goddard is an extension professor of medical and veterinary entomology at Mississippi State University, where he speaks, writes, and conducts research on a wide variety of medically important pests. In addition to his extension duties, Dr. Goddard teaches courses in medical entomology and forensic entomology and mentors a number of graduate students. Prior to coming to MSU, Dr. Goddard was the state medical entomologist at the Mississippi Department of Health, where, for 20 years, he developed and implemented vector control projects throughout the state of Mississippi. Dr. Goddard has written well over 200 scientific papers, 10 book chapters, 4 college-level reference or textbooks, and 8 fiction novels. He and his wife, Rosella, live in Starkville, Mississippi, have two grown sons, and four grandchildren.

section one

---

*Essentials of Public Health
and Entomology*

chapter one

History of Medical Entomology and Public Health

Introduction and Background

"The more things change, the more they stay the same" is an appropriate quote for public health entomology in the 21st century. If one reads some of the oldest works on medical entomology and preventive medicine, for example, Griscom's speech before the New York Academy of Sciences (1855),[1] the *Bulletin of the Sanitary Commission* (1863),[2] Doan's *Insects and Disease* (1910),[3] and Pierce's *Sanitary Entomology* (1921),[4] the themes are exactly the same as those found in entomological, medical, and public health literature today:

1. Entomology bears a two-fold relationship with human health: helping provide adequate food supply and preventing disease transmission.
2. Disease agents can be transferred to humans by several methods, including direct contact, food, insects, soil, and fomites.
3. Both mechanical and biological transmission of disease agents by arthropods are important, as well as efforts to block them.
4. All attempts to link arthropod transmission or causation of disease need to be rigorous and well thought out.
5. Basic sanitation measures such as clean water, sewage disposal, and use of screen wire for windows and doors are absolutely critical in preventing vector-borne disease. In fact, sanitation is the superior activity in disease control, even above quarantine.[5]

Even in the midst of modern technology, such as complicated molecular tools for identifying vector-borne diseases and their agents, these above-mentioned themes should be overarching guides to our research, education, and disease prevention efforts. Public health is often a matter of doing the same thing over and over—hammering home certain unalterable facts about health, disease prevention, and ways to remain healthy. Further, the short nearly 150-year-old era of modern medical entomology has taught us to be prepared, to be ready, for disease outbreaks (both old and new ones) resulting from disasters and wars (for a good discussion of medical entomology during wartime, see Cushing).[6]

3

Medical Entomology versus Public Health Entomology

Medical entomology is the scientific discipline of the study of insects, but often includes other arthropods, which may directly or indirectly affect human health. Negative effects from arthropods may range from blisters, bites, and stings, to disease transmission and allergic reactions. Medical entomology is largely an academic discipline housed in university entomology or biological sciences departments.

Applied aspects of medical entomology may be seen in military units, focused on protecting military personnel from arthropod-borne diseases, or federal, state, or local governmental agencies such as state health departments or the U.S. Centers for Disease Control and Prevention. This is where medical entomology becomes public health entomology—applied entomology pertaining to human health, safety, and well-being.

Public health entomology is concerned with not only the arthropods themselves, but also the prevention, surveillance, and control of vector-borne diseases. More importantly, protection against pests and vector-borne diseases includes an enforcement or regulatory function, which sets public health entomology apart from traditional medical entomology. Reviewing the history of medical entomology reveals the distinctions between academic and applied entomology and the importance of applied entomology to the field of public health.

Historical Aspects of Medical Entomology

Long before anyone knew about the causative agents of medical conditions, it was recognized that insects might produce diseases.[4] About 2500 BC, a Sumerian doctor inscribed on a clay tablet a prescription for the use of sulfur in the treatment of itch, a chemical we now know kills itch and chigger mites.[6] Other recorded instances of arthropod-borne diseases and infestations can be found in the Old Testament, beginning with accounts of plagues on the Egyptians. For example, the third plague, called "lice" in the King James version of the Bible (in later translations more accurately termed "gnats"), was likely *Culicoides* midges, which transmitted the causative agents of African horse sickness and bluetongue to Egyptian livestock (which was, by the way, the fifth plague).[7] The sixth plague consisted of boils and ulcers on humans and animals, which could have been the disease called Glanders, transmitted by biting flies.[7] More recent evidence of health issues from arthropods has been found as well. First-century BC hair combs containing remains of lice and their eggs have been unearthed in the Middle East.[8] Peruvian pottery from about 600 AD shows natives examining their feet—and their feet display what appear to be holes where chigoe fleas (burrowing fleas) had been removed.[9,10] Other pottery found near the Mimbres River, New Mexico, dated to 1200 AD, clearly depicts a swarm of mosquitoes poised for attack. For hundreds of years after that, there were hints and suggestions made by various people who imagined a connection between insects and diseases. For example, in 1853, Dr. Louis Beauperthuy, a French physician, elaborately argued that yellow fever is transmitted to humans by mosquitoes (but he thought it was by mechanical transmission from decomposing matter which they had visited).[11] But these suggestions were almost completely ignored by the medical community. In 1871, the idea that any specific disease might be insect-borne was not even mentioned in any of the standard medical literature.[11]

That started to change in the late 1800s, when several fundamental discoveries were made over a 20-year period linking arthropods with the causal agents of disease.[11] In 1878, Patrick Manson observed development of the nematode *Wuchereria bancrofti* in the body of the mosquito *Culex quinquefasciatus*, and eventually he and others proved that mosquitoes were indeed the vector of these filarial worms. Charles Laveran, in 1880, found that a protozoan may be the causative agent of human malaria, and 9 years later, Theobald Smith discovered the protozoan *Babesia bigemina*, causative agent of Texas cattle fever. In 1893, Smith and F.L. Kilbourne proved that the cattle tick *Boophilus annulatus* is the vector of Texas cattle fever. Two years later, David Bruce investigated the animal disease nagana and found that its vector is the tsetse fly. In 1897, Ronald Ross linked malaria parasites[12] to certain mosquitoes, which he and others later identified as *Anopheles*. Walter Reed and his colleagues first reported an association between mosquitoes and yellow fever,[13,14] while a few years later, Carlos Finlay and other members of the Yellow Fever Commission definitively proved that yellow fever is carried by the mosquito *Aedes aegypti*. In 1902, David Bruce discovered the causative agent and vector of African sleeping sickness (trypanosomiasis), and in 1909, Nicolle established that the body louse is the vector of *Rickettsia prowazekii*, the agent of epidemic typhus.[15] Although these are the primary (early) discoveries in medical entomology, the list goes on of modern major advancements concerning arthropods and the role they play in disease transmission.

Arthropods themselves, as well as diseases they transmit, have greatly influenced human civilization. Sometimes the influence has been notable or recorded, such as when plague epidemics swept through the Middle East or Europe, louse-borne typhus decimated armies, or yellow fever destroyed entire armies or cities. There is an account of plague in Egypt circa 1200 AD, stating that more than a million people died.[16] The famous French emperor Napoleon Bonaparte crossed the Niemen River into Russia in June 1812 with 420,000 men, but within 6 months, he only had 3,400 men left, most having died from epidemic typhus.[16] Typhus also ravaged Russia from 1918 through 1922, leading to approximately

30 million deaths in the civilian population.[17] Yellow fever has been just as devastating. During the Haitian–French War (1801–1803), Napoleon's largest expeditionary force of approximately 50,000 soldiers was almost completely destroyed by yellow fever.[18] An almost total disintegration of society is described by Crosby in her book about the yellow fever epidemic in Memphis, Tennessee.[19] The last epidemic of yellow fever in this country occurred in 1905 in Louisiana, with 9,321 cases and 988 deaths.[20]

In other instances, the influence of arthropods has not been easily recognized. Great expanses of land near seacoasts (e.g., Florida) or inland swamp areas were left undeveloped because of fierce and unbearable mosquito populations. These areas were only populated after the advent of effective area-wide mosquito control. In a similar fashion, large parts of Africa were left untouched by humans for centuries because of African trypanosomiasis (sleeping sickness) and falciparum malaria.

Malaria as an Example of Historic Public Health Entomology

Certainly, prehistoric humans were subject to malaria. There is archeological evidence of human malaria from the eastern Mediterranean region as far back as the Neolithic period (up to 12,000 years ago) and possibly also in Southeast Asia at that same time.[21] Most likely the disease originated in Africa, but subsequently followed human migrations to the Mediterranean shores, to Mesopotamia, the Indian peninsula, and Southeast Asia.[22] There are many ancient references to seasonal fevers in Assyrian, Chinese, and Indian religious and medical texts which could have been descriptions of malaria. At that time, little was known about the disease or what to do about it, other than use of incantations or sacrificial offerings. It wasn't until the 5th century BCE that the Greek physician, Hippocrates, discarded superstition for logical observation of the relationship between appearance of the disease and seasons of the year and places where his patients lived.[22] He was the first to describe in detail clinical malaria and its complications. The Greeks began to associate seasonal fevers with stagnant waters and swamps, and wrongly thought it could be contracted by drinking swamp water.[23] Later, the Romans believed the disease to be caused by breathing "miasmas" or vapors from stagnant water, and thus began to call it, "bad air" or mal-airia (malaria).[23]

For the next 1,500 years or so no new information or treatments concerning malaria were forthcoming. However, early in the 17th century, physicians began to use 'Peruvian bark' for treatment of all types of fevers, some of which responded quite well to the treatment. In 1735, Linnaeus named members of that genus of flowering plants *Cinchona*, which contained the chemical quinine, a treatment for malaria (but the actual active ingredient was not isolated until 1820 by Pelletier and Caventou in France in 1820).[22] The vector of malaria was shown by Ronald Ross in 1897 when

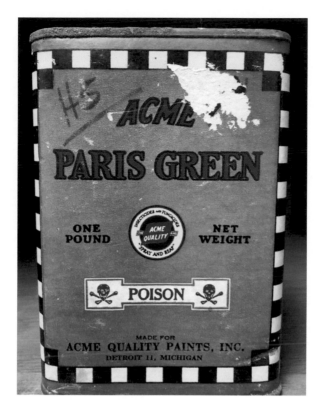

Figure 1.1 Paris green insecticide, which contained arsenic, lime, and copper.

Source: Photo copyright 2021 by Jerome Goddard, Ph.D.

he found the developing form of malaria parasites in the body of a mosquito which had previously fed on a patient with the plasmodia in his blood.[12] During the first and second world wars, much progress was made in synthesizing new antimalarial drugs such as mepacrine and chloroquine,[22] while other methods of malaria control were also investigated, using Paris green for larvicides (Figure 1.1) and pyrethrum sprays in dwellings.[24]

Malaria in Mississippi

In the United States, the fight against malaria was particularly intense in the southern states (Figure 1.2). In the early 1900s, the American South experienced a peak of malaria transmission, while the disease had been mostly eradicated north of Ohio.[25] The South had only one-third of the U.S. population in 1940, but 96% of all reported malaria deaths.[26] In 1941,

Figure 1.2 Malaria research worker in the laboratory.

Source: Photo courtesy National Library of Medicine

the malaria rate for the United States as a whole was 0.1 per 100,000 population, while that for the 14 southern states was 2.73.[27] Heavily endemic malaria foci included the delta area of the lower Mississippi Valley from about Cairo, Illinois, to Natchez, Mississippi (Figure 1.3). In fact, one study of Sunflower and Bolivar Counties in Mississippi from 1916 to 1918 revealed a staggering 50% prevalence of malaria among the population.[28] A large portion of deaths occurring in the Mississippi Delta region at that time was due to malaria, with hemoglobinuria (bloody urine) and so-called congestive chills being the most commonly reported causes of death.[29]

There are at least four malaria vectors occurring in the United States—*Anopheles freeborni* (western United States), *An. hermsi* (a fairly recently described species in the western United States), *An. punctipennis* (eastern and western United States), and *An. quadrimaculatus* (eastern United States). *Anopheles quadrimaculatus* is a complex of five nearly identical species, although one of them, *An. quadrimaculatus* s.s., was historically the primary species involved in malaria transmission in the southeastern United States (Figure 1.4).[30] In 1929, Manuel Perez performed an anopheline survey of the state of Mississippi and found that *An. quadrimaculatus*,

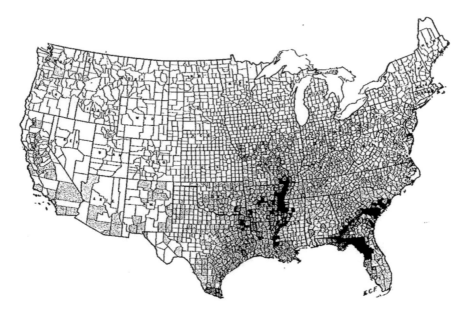

Figure 1.3 Average reported malaria deaths by counties. *Source*: Redrawn from Faust, 1941

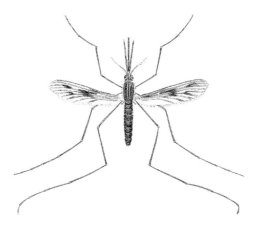

Figure 1.4 *Anopheles quadrimaculatus*, the primary vector of malaria in the SE U.S. *Source*: Figure from the Public Health Service

An. punctipennis, and *An. crucians* (now recognized as a species complex)[31] were distributed statewide.[32] Figure 1.5 shows a map of Mississippi with the physiographic regions of the state labeled, including the main ones—the delta, the north-central plateau, the Tennessee-Tombigbee hills, the

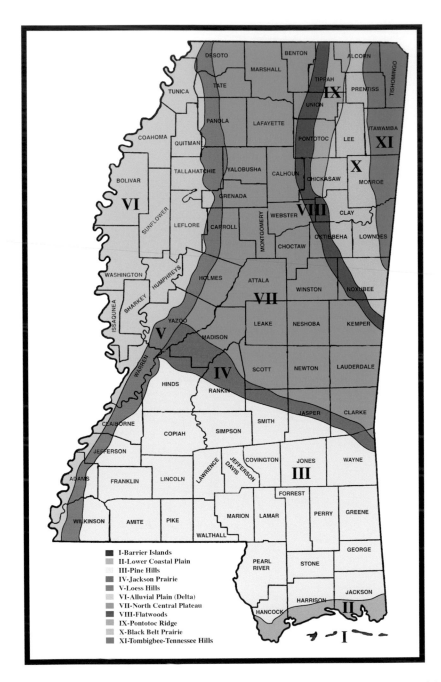

Figure 1.5 Major physiographic regions of Mississippi. Note: *Anopheles quadri-maculatus* are most common in the delta and north-central plateau, where slow-moving water sources such as lakes, streams, and ponds are most prominent. *Anopheles punctipennis* and *An. crucians* are most common in the north-central plateau and pine hills region. *Source*: Adapted from Lowe EN, *Mississippi State Geological Society Survey Bulletin*, 14, 1919:1–346

pine hills (piney woods), and the coastal plain. Other researchers found that *Anopheles quadrimaculatus* was most common in the delta and north-central plateau where slow-moving water sources such as lakes, streams, and ponds were located.[33] On the other hand, *Anopheles punctipennis* and *An. crucians* were most common in the north-central plateau and pine hills region. These data demonstrated that the distribution of malaria and *An. quadrimaculatus* overlapped in the South, leading the authors to conclude that this species was responsible for the endemic presence of malaria in that region.[34]

Regional Differences/Patterns of Outbreaks

In the first half of the 20th century malaria was reported from every county in Mississippi.[35] In 1913, return-postage-paid postcards were sent to physicians in all 79 counties within the state inquiring about the incidence of malaria within their county. In August of that year, 2,009 cards were mailed and 528 were returned representing all 79 counties. Results showed that there were 14,753 cases diagnosed in the month of August, and 1,142 of them were confirmed microscopically. Interestingly, of the 528 cards returned, only 136 physicians (representing 50 counties) reported using a microscope to diagnose malaria. This translates to only 26% of physicians using microscopes for diagnosis of malaria during this time.[35] This might explain inaccuracies encountered when looking at the incidence of disease among different sources. In 1946, there were approximately 17,000 cases of malaria reported to the Mississippi State Board of Health. However, in the following year less than 1,000 cases were reported. The reason for this drastic difference was due to the state now requiring that the pathogen be identified in each case and including a $5.00 bounty for each confirmed case.[36] Diagnostic techniques prior to 1947 probably resulted in inaccurately diagnosed cases, and thus these inaccurate statistics.

Complicating the situation, malaria was likely blamed for illnesses that it did not cause.[37] In autumn, when a large number of people came down with symptoms of fever and chills, many physicians believed that a microscope was not needed to diagnose malaria, and usually they just treated with quinine. A country physician gives his account of suspect malaria cases and their treatment:

> We had ague as a regular disease, and it was not difficult to diagnose. You could feel it with the naked eye. Other people could also feel it when that patient had a chill, for he shook the house. Our standard remedies for ague were calomel, castor oil, and quinine, and they were not measured out on the apothecary's scales.[38]

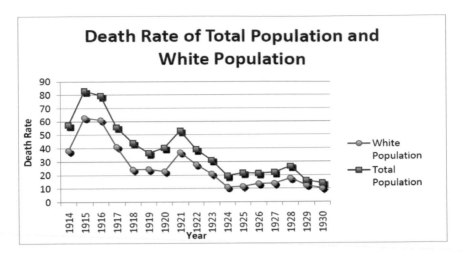

Figure 1.6 Death rates from malaria in Mississippi, whites vs. non-whites.

Furthermore, many people likely never sought medical treatment for malaria and lived with chronic infection for most of their lives. These people would also never have been reported as "cases" and would have contributed to the inaccuracy of malaria incidence reports.

The incidence of malaria also varied by race and various regions of the state. In terms of overall numbers, the disease affected whites more than nonwhites, but this was thought to be a result of better reporting. Blacks were generally refractive to *P. vivax* infection, but susceptible to *P. falciparum*. As for death rates, nonwhites had decidedly higher numbers (Figure 1.6). For example, in 1930, the white death rate from malaria in Mississippi was 10.0 per 100,000, while the black death rate was 17.8.[34] As mentioned, the highest incidence of malaria in Mississippi occurred in the delta region (with the largest population of blacks) followed by the bluff, northeastern, southeastern, and coastal regions.[35]

Factors Contributing to Malaria in Mississippi

Different regions of Mississippi had different factors influencing incidence and prevalence of malaria that had to be considered for prevention and control purposes. For example, in many parts of the world, yearly malaria mortality rates correspond to annual rainfall amounts. This was not the case for Mississippi historically. Figure 1.7 shows the average rainfall compared to the total number of malaria deaths within the state. As annual rainfall increased, malaria deaths often decreased and vice versa. Thus, annual rainfall and temperature were apparently not important factors influencing the incidence of malaria in the United States.

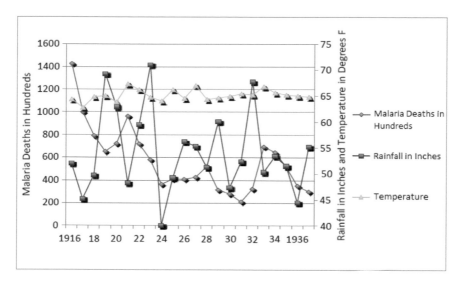

Figure 1.7 Malaria deaths in hundreds versus the average temperature and rainfall within Mississippi.

Socioeconomic factors better explained the historical incidence of malaria in Mississippi.[39] Disease prevalence was much higher in rural districts than in the larger cities and was thus considered a rural disease. Rural areas had houses that were inadequate with occupants too poor to make improvements to the dwelling. Areas with the most malaria had houses not owned by the occupants and with occupants who did not remain in the same place very long.[40] Further, poor people without access to electric fans often sat outside on the porch during hot weather, exposing themselves to malaria-carrying mosquitoes (Figure 1.8). These were some of many reasons why malaria was so difficult to eradicate.[41]

During the 1900s, the livelihood of people living in Mississippi was heavily dependent upon agriculture, with the primary crop being cotton. In the delta, there were large numbers of blacks and whites who occupied poorly constructed tenant "cabins," often situated near swamps and bayous. Further, with the state being so dependent upon the cotton crop, a majority of the people were left with an inadequate food supply when cotton prices were low.[42] This suggests that income played a significant role in the health of farmers and their tenants. Income as reported on tax returns is a good indication of the economic condition of the people. Figure 1.9 shows a strong inverse relationship between the tax return data and malaria deaths over the course of the 22 years, except for 1929 to 1931. As personal income decreased during the depression, the number of deaths from malaria increased and vice versa. Theoretically, during years

Figure 1.8 People sitting outside were often exposed to malaria-carrying mosquitoes.

Source: Photo courtesy National Library of Medicine

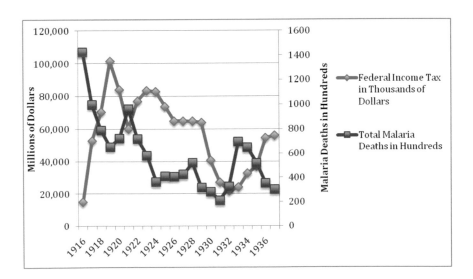

Figure 1.9 The net income tax return versus malaria deaths in Mississippi.

of high income, people could afford better protection and treatment from the disease, such as better clothing, quality food for nourishment, proper screening on houses, medical services, and therapeutic drugs, all of which are important in the health and immunity of people trying to ward off disease.

Pesticides and Entomologists in the Fight against Malaria in Mississippi

By 1900, 40 of the 45 states had established health departments. The first county health departments were formed in 1908,[43] and in 1929, Dr. Felix Underwood, state health officer of Mississippi, initiated efforts to make malaria control an integral function of local (county) health departments.[44] He believed that temporary or transient disease control workers sent from the federal government were important, but not near as much as trained local personnel who he said "have not only the basic training, but that indefinable public health viewpoint and sufficient knowledge of economics and sociology peculiar to the territory involved." This was among the first efforts nationwide to establish trained sanitarians in every county to investigate diseases, perform inspections, and conduct general sanitation education and investigations. Within the next decade, the U.S. Public Health Service established malaria control units within health departments of several southern states so that control efforts could be more organized and systematic.[45] The idea was to switch health department malaria control efforts from passive to active—finding and treating every case of malaria.[46] Entomologists were also assigned to these units and their duties included basic research, surveillance, and control efforts. Control of malaria seemed to be divided among those who wanted to destroy the vector and those who wanted to treat the parasite within the human host.[47] Control efforts consisted of pesticide use (such as with Paris green) (Figure 1.10), destruction of breeding sites, and promotion of screen wire for windows (Figure 1.11). Screening played a substantial role in malaria prevention. For example, in 1905, approximately 5% of houses in Sharkey County, Mississippi, had screen doors and windows. By 1931, this number had increased to 64%.[48] Control efforts aimed at the parasite included better distribution and use of specific antimalarial drugs such as quinine (Figure 1.12), as opposed to a wide variety of "snake oil" products advertised to cure any number of medical conditions. These products contained mostly alcohol, but also opium and chloroform (Figure 1.13). Lastly, the malaria control included efforts to improve socioeconomic conditions.

Until 1945, many efforts were implemented that significantly reduced the morbidity and mortality of malaria but never completely eliminated

Figure 1.10 Larviciding for mosquitoes using an airplane.

Source: TVA photograph

Figure 1.11 Proper screening of houses was important for malaria control.

Source: US Navy photo

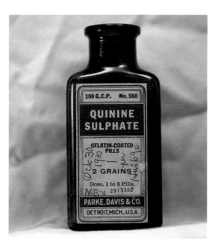

Figure 1.12 Quinine tablets were the primary treatment for malaria in the United States in the early and mid-20th Century.

Source: Armed Forces Pest Management Board photo by David Hill

Figure 1.13 Some commonly available medicines were inefficient against diseases, sometime even harmful. This product contained alcohol, opium, and chloroform.

Source: Photo copyright 2021 by Jerome Goddard, Ph.D.

the disease. It is difficult to say for sure which had the largest effect because all may have played some role in the collaborative eradication of the disease from the United States. One likely key factor was the switch from passive to active malaria surveillance with an effort to find and treat every case. Interestingly, during times when malaria had decreased or disappeared, there were still large numbers of *Anopheles* mosquitoes present. This seems to be a factor unrelated to the presence of malaria.[49] In addition, improved case reporting and research data better identified problem areas and locations where control efforts could be focused. Housing for livestock was also improved, which redirected mosquitoes away from humans. The quality of human dwellings had also improved by the mid-1940s, and increased personal income allowed for repairs to screening on houses. Education was also an important factor that led to prompt recognition and treatment of cases as well as use of screening. The pesticide DDT became widely available for mosquito control in the South after 1945.[50] Finally, medical treatment and quinine became more affordable and accessible to citizens. For example, quinine sold for about $4.00 per ounce in 1896, but that fell to about 25 cents per ounce in 1913.[49] Barber states, "The factors concerned in the diminution of malaria in the United States are interdependent; their importance has varied with time and locality, but all have been closely related to the agricultural development of the country."

Although the public health entomological efforts were herculean, removal of malaria in the southern United States was probably more likely related to development and enhancement of rural lifestyles than campaigns by health agencies. Improvements of social conditions diminished the malaria threat in spite of the large numbers of *Anopheles* still present. The last case of locally acquired malaria within Mississippi occurred in 1955, with the last death occurring in 1953. Since then, mosquito control efforts have been mainly directed toward practical or nuisance mosquito control, instead of malaria prevention and control. While this account of applied entomology to abate malaria in the South is a great example of public health entomology, the science of public health actually had its origins much earlier (see section on early beginnings of public health). Interestingly, malaria control in the United States led to establishment of the Centers for Disease Control.

Inception of the Centers for Disease Control

Founded in 1946 as a branch of the United States Public Health Service, the (now named) Centers for Disease Control and Prevention (CDC) was originally created to continue the work of the World War II Malaria Control in War Areas program. The primary focus of the agency was preventing the spread of malaria in the United States. Because the South was known as the heart of the malaria zone in the United States (see the previous section), the first offices of what was then called the Communicable Disease Center were established in Atlanta, GA on a single floor of a downtown office building. Less than 400 employees staffed the Communicable Disease Center, many of them entomologists who spent their workdays applying pesticides and performing habitat abatement measures to eradicate *Anopheles* mosquito populations. Apropos, their first organizational chart is often presented as a mosquito figure (Figure 1.14). After success of the CDC's malaria control program, the

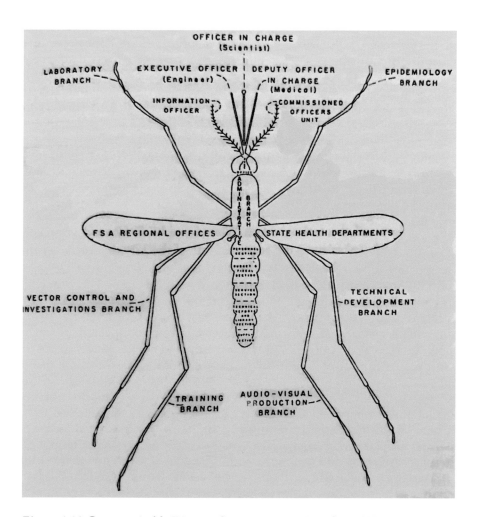

Figure 1.14 Communicable Disease Center organization chart, 1951.

agency's founder, Dr. Joseph Mountin, urged the federal government to expand the role of the CDC to focus on surveillance and control of other communicable diseases. The U.S. government recognized the value of monitoring communicable diseases, and therefore the agency expanded its surveillance activities with establishment in 1951 of the Epidemic Intelligence Service (EIS), a post-doctoral program designed for epidemiological training. In subsequent decades, EIS would further expand to investigate rabies, polio, anthrax, influenza, cholera, and smallpox, among many other severe illnesses.

As disease surveillance and control activities continued, the CDC experienced several official name changes; to the National Communicable Disease Center in 1967 and subsequently to the Center for Disease Control in 1970. It wasn't until 1992 that the agency's name was officially changed to the Centers for Disease Control and Prevention as part of the Preventive Health Amendments of 1992. While the acronym for the agency remained "CDC" due to its recognizability within the public health community, this name change was enacted to acknowledge the CDC's profound role in the prevention of illness, injury, and disability in the United States. This forward-thinking approach to the understanding and prevention of disease remains a primary focus for the CDC in the 21st century, with the agency now supporting over 10,000 full-time federal employees from a wide range of disciplines. The main CDC campus now occupies 15 acres of land on Clifton Road in Atlanta, with multiple field offices across the United States.

Chapter 8 of this book (*Where to Go for Help*) provides more information about the various Centers within the Centers for Disease Control, but two CDC divisions are perhaps most applicable to the scope and aims of this book—the Division of Vector-borne Diseases and the Division of Parasitic Diseases and Malaria (Table 1.1). Branches of these two divisions are mostly split between the main CDC campus in Atlanta, Georgia and the Ft. Collins, Colorado site. In the 1940s, personnel associated with field station laboratories located in Cache Valley, Utah and Greeley, Colorado conducted surveillance, research, and prevention of mosquito-borne diseases that cause swelling of the brain (encephalitis). In 1967, these laboratories consolidated on the Colorado State University (CSU) campus in Fort Collins to enhance collaboration between CSU and CDC scientists. In the 1960s, a CDC branch dedicated to plague moved from San Francisco, California to Fort Collins when the majority of plague cases were occurring in New Mexico, Arizona, and Colorado. In 1989, Lyme disease, now part of the Bacterial Diseases Branch, was moved to Ft. Collins. Today, the Fort Collins site is the only major CDC infectious disease laboratory outside of Atlanta, Georgia.

Table 1.1 Key Divisions for Vector-borne Diseases at the Centers
for Disease Control.

CDC Division or Branch	Diseases Include	Location
Division of Vector-Borne Diseases		
• Arboviral Diseases Branch	West Nile, Chikungunya, Zika, Yellow fever	Ft. Collins, CO
• Bacterial Diseases Branch	Lyme disease, Plague, Tularemia	Ft. Collins, CO
• Dengue Branch	Dengue	San Juan, Puerto Rico
• Rickettsial Zoonoses Branch	Rocky Mountain spotted fever, Q fever, Typhus fevers	Atlanta, GA
Division of Parasitic Diseases and Malaria		
• Malaria Branch	Malaria	Atlanta, GA
• Parasitic Diseases Branch	Chagas' disease, Lymphatic filariasis, Onchocerciasis	Atlanta, GA
• Entomology Branch		Atlanta, GA

Military Medical Entomology

Military entomology was almost unheard of in 1870, as mosquitoes, flies, lice, ticks, and fleas were regarded as mere nuisances. That began to change after 1898 during the 118-day Spanish–American War where there were 369 battle casualties among U.S. troops, compared to 1,939 soldiers who died from diseases, primarily typhoid fever.[51] A Board of Army Medical Officers concluded that spread of this disease was due to flies. During World War I, military leaders understood too well that casualties from arthropod-borne disease exceeded those from bullets and bombs, especially trench fever, a newly recognized illness transmitted by body lice. For the first time, approximately eight entomologists were commissioned in the Army to assist in control of insects (primarily) in Army camps.[51] Interestingly, there is a fascinating story of a female concert pianist, Clara Ludlow, who studied music at the New England Conservatory of Music in Boston, MA, but eventually abandoned that for her (real) passion in science and enrolled at the all-male Agricultural and Mechanical College of Mississippi (now Mississippi State University), where she got a B.S. in Agriculture and then an M.A. in Botany.[52] She went on to become a major influence in military medical entomology, publishing 49 scientific papers and describing 72 new mosquito species and 6 genera. Clara was the first woman elected to be an active member of the American Society of Tropical Medicine, and was ultimately buried in Arlington National Cemetery in recognition for her service to the country. Some people consider her to be the first "military entomologist" since she performed mosquito taxonomy for the U.S. Army from 1901 until her death in 1924.[51,52]

World War II again proved that arthropods exact a heavy toll on soldiers during wars. Harvey Schultz[1] said, "It seems predictable today that a little dapple-winged mosquito, the *Anopheles*, would cause five times more casualties among U.S. troops during WWII than would battlefield injuries."[51] General MacArthur is said to have stated in 1943, "This will be a long war, if for every division I have fighting the enemy, I must count on a second division in the hospital with malaria, and a third division convalescing from this debilitating disease."[51] Time lost by Army combat troops during WWII because of malaria has been estimated at 10,140,872 soldier days.[6] Malaria wasn't the only vector-borne disease adversely affecting U.S. troops. Dengue, scrub typhus, Q fever, and louse-borne (epidemic) typhus also decreased unit strength by 10–30%.[51] There is a description by Warrant Officer E. J. Kahn, Jr. of American soldiers in the early days of WWII in the Pacific theater:[53]

> The men at the war front in New Guinea were perhaps the most wretched looking soldiers ever to wear the American uniform. They were gaunt and thin, with deep black circles under their sunken eyes. They were covered with tropical sores. . . . They were clothed in tattered and stained jackets and pants. . . . Often, the soles of their shoes had been sucked off by the tenacious, stinking mud. Many of them fought for days with fevers and didn't know it. . . . Malaria, dengue fever, dysentery, and in a few cases typhus hit man after man. There was hardly a soldier among the thousands who went into the jungle who did not come down with some kind of fever at least once.

Eventually, the military began to organize preventive medicine and malaria control units for deployment in war zones, as well as intensive educational efforts about malaria (Figure 1.15). By 1944, the War and Navy Departments had placed 771 specially trained personnel in the field. Army entomologists staffed 17 malaria survey and control detachments. The Navy Division of Preventive Medicine established epidemiology units, which by 1944 numbered 122. Over 100 malaria control units led by entomologists and medical officers were distributed throughout Naval units in forward areas. Navy entomologists were also deployed to China, North Africa, the Caribbean, and Central Africa.[51] The need for

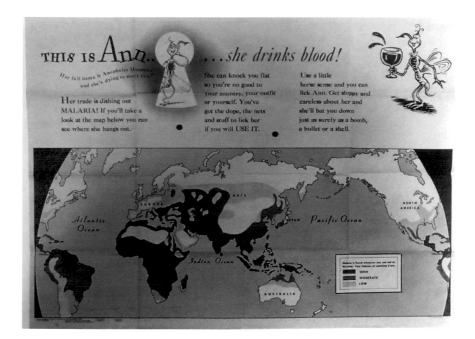

Figure 1.15 The U.S. Army utilized this figure, drawn by Dr. Seuss, in its malaria educational efforts.

Source: Figure courtesy National Library of Medicine, public domain

widespread spraying of insecticides for malaria control became critical. The Army and Navy pursued spraying technology using helicopters, while the Army Air Force (AAF) perfected aerial spraying using fixed wing aircraft, equipping L-4 aircraft for aerial spraying in New Guinea in 1944.[51] These techniques and equipment were eventually modified for B-25 and C-47 airplanes. After the end of the war, a unit called the Special DDT Flight was created, but was soon transformed to the Special Aerial Spray Flight (SASF) in 1947 when the Air Force became a separate armed service. After more than 25 years at Langley Air Force Base, Virginia, the SASF was transferred from the active Air Force to the Air Force Reserve in 1973, but prior to this transfer, the SASF had sprayed for mosquitoes, Japanese beetles, and fire ants in various locations at the request of the Army, Navy, and other federal agencies. Relocated to Rickenbacker Air Force Base, Ohio, the unit was soon renamed the Spray Branch of the 907th Tactical Airlift Wing. In 1986, the Spray Branch began to transition from C-123 airplanes to C-130A airplanes and developed the modular aerial spray system for use in C-130E and H airplanes. In 1991, the aerial spray mission was moved to the Youngstown Air Reserve Station, Ohio (see Chapter 6, *Military Aerial Spraying Capability*).

Arthropod-borne diseases also played a significant role during the Korean and Vietnam wars. Malaria, Japanese B encephalitis, relapsing fever, typhus, and dysentery cases were widespread. Entomologists sprayed for mosquitoes and flies, and also helped with delousing efforts using DDT, and eventually lindane. In Vietnam, at the height of the conflict, disease was responsible for over 70% of Army hospitalizations, of which malaria and fevers of unknown origin accounted for half.[51] In many cases, the problem from malaria during the Vietnam war was not from lack of spraying and other environmental control efforts, but instead, from soldiers purposely not using personal protection techniques against mosquitoes and taking their anti-malaria pills. Therefore, entomologists initiated educational campaigns concerning personal protective methods such as proper wearing of the uniform and use of bed nets and repellents.

Infectious diseases transmitted by arthropods impacted American troops during Operations Desert Shield/Storm, but not as much as had been predicted.[54] There were only seven cases of malaria, three cases of Q fever, and one case of West Nile fever identified among U.S. troops.[55] One major concern that failed to materialize was the threat of sand fly fever and leishmaniasis transmitted by sand flies (see Chapter 13).[54] Sand flies were indeed a biting pest during the conflict, but cases of cutaneous and visceral leishmaniasis in U.S. soldiers were few, if any. Low numbers of illnesses caused by sand flies and other insects were attributed to:[54]

1. Deployment of troops to mostly barren desert locations.
2. Deployment of troops during cooler winter months (December–February).
3. Widespread use of insecticides and repellents.

Currently, the U.S. military has well over 100 medical entomologists. The Army has 60 active duty uniformed entomologists, the Navy has 46, and the Air Force has 17. All the Air Force entomologists are now Public Health Officers who carry a Specialty Shredout identifying them as medical entomologists. Although originally aimed at surveillance, prevention, and control of vector-borne diseases (Figure 1.16), military entomologists have since broadened their mission to include providing equipment, expertise, and pesticide advice for nuisance pests, relieving human suffering during disasters, and re-establishment of normal living conditions afterwards.[56] Further, their duties may now entail bird/aircraft strike hazard reduction, weed control, and wood protection.[51]

Figure 1.16 The author was once an Air Force medical entomologist (1986–1989), responsible for arthropod vector surveillance, identification, and control.

Source: U.S. Air Force Photo

Early Beginnings of Public Health

In the very beginnings of written history we find that health and hygiene issues were assigned to priests and other religious leaders. Almost 5,000 years ago, there was a position in Egyptian government called the *vizier*, a priestly minister of state who had the duty of inspecting the water supply every 10 days. The priesthood of the biblical Hebrews was similar, albeit more sophisticated and elaborate. Although its origin may have been based in part on Egyptian and Mesopotamian medical knowledge, the Mosaic Code developed by the Hebrews is the basis of modern sanitation and pest control today. This code was the best means of achieving health devised by humans for nearly 3,000 years,[57] and included insights such as establishing quarantines, washing hands with running water, limiting contact with the dead, and eating meat only from healthy animals slaughtered under close scrutiny of concerned rabbis. Again, responsibility for carrying out the regulations of the Mosaic Code rested with the priesthood. Later, in classical Greece, science, medicine, philosophy, and the arts flourished for almost 500 years. Asclepius was the patron saint of Greek medicine and his daughter, Hygeia (Figure 1.17), is considered the

HYGEIA. "You doubtless think that as all this filth is lying out in the back streets, it is no concern of yours. But you are mistaken. You will find it stealing into your house very soon, if you don't take care."

Figure 1.17 Hygeia was the patron saint of preventive medicine.

Source: Photo courtesy National Library of Medicine

saint of preventive medicine (some health departments still use the image of Hygeia as their logo). During the Roman Empire, Galen's medical writings became the authoritative source for all Arabic and late medieval medicine.[58] The Romans made great strides in sanitation, civil engineering, and hygiene, with construction of numerous aqueducts and sewage systems. They were also among the first to develop baths and indoor sanitation measures. The Roman position of *aediles* (about 500 BC) was essentially a nonphysician administrator of sanitation, and included the responsibility to care for the safety and well-being of the city of Rome, specifically, repair and upkeep of public buildings, aqueducts and sewers, fire prevention, and street improvements. After the Roman Empire, and during the ensuing Dark Ages, science basically disappeared from Europe, and along with it, much of the sanitation and health-related arts. There is some evidence that monks and other church leaders rarely bathed, considered cleanliness as an abhorrence, and, at least according to Snetsinger, were vermin-ridden. Plague, leprosy, smallpox, and tuberculosis emerged among the population, decimating Europe from about 800 AD to 1400 AD. However, there were still remnants of public health and sanitation administration. For example, in England in 1297, there were legal regulations requiring every man to keep the front of his tenement clean, and in 1357, a royal order was issued by King Edward III requiring the mayor of London and the sheriffs to prevent pollution of the Thames River.[20] King John of France in 1350 established "sanitary police" as a governmental entity.[20] Ever so gradually, cleanliness and hygiene returned as important factors in society. In 1491, Johann Pruss promoted regular bathing and changes of clothing as a lice deterrent, and Ignatius Loyola, the famous Catholic priest, established rules of exercise, nutrition, and cleanliness for his followers. In the late 1500s, Queen Elizabeth I instituted housing regulations proposed to relieve overcrowding, pure food laws to prevent adulteration of foodstuffs, and regulations for the control of epidemics. During this time, pests became recognized as being related to filth, waste, and harborage. Thomas Tryon in 1682 reported that vermin are encouraged by filth and can be avoided by washing and exposure to fresh air. The Renaissance opened the door for even more improvements in sanitation and hygiene in the human population that have continued to this day.

ANCIENT DEVELOPMENTS IN PUBLIC HEALTH AND HYGIENE

- Establishment of Egyptian *viziers* (approximately 3000 BC), who inspected the water supply regularly to make sure it was safe.
- The Hebrew Mosaic Code, which included rules for quarantines, personal hygiene, and close scrutiny of animal slaughter methods.
- Classical Greece, rapid development of the disciplines of science and medicine.
- Roman Empire, great progress in civil engineering, clean water supply, and sewage disposal.

Public Health in General

The science of public health is focused on protecting the health of entire populations rather than individuals. Dr. Mark Boyd[59] provided an excellent treatise of this issue and administration provided by public health officers. The following sections are patterned after his description of the role(s) of a State Health Officer. Public health practitioners are concerned with protecting and improving the health of communities through education, promotion of healthy lifestyles, regulation of health-related activities and institutions, and research into community disease and injury prevention. These populations can be as small as a local neighborhood or as big as an entire country or region. On the other hand, private practice physicians (clinicians) are oriented toward individual people and their illnesses. Their practice centers on isolated, often unrelated, individual maladies. Private practice compels physicians to give each patient individual consideration. Public health workers facilitate, or are required to receive information from, the health community in order to see the big picture. By cooperating with each other, clinical medicine and public health can help each other improve health maximally, while emphasizing society's responsibility to promote both healthy environments and consistent, high-quality care.[60]

Public health professionals focus on population-level–based services and try to prevent problems from happening or recurring through surveillance of diseases and conditions, implementing educational programs, regulating health systems and professions, developing policies, administering services, and conducting research (Figure 1.18). One of the most fundamental aspects of public health is surveillance, which indicates a baseline level of health in the population and detects changes, both positive and negative, as indicators of health status (see Chapter 4). Simply, surveillance allows public health officials to know what's out there. Surveillance for diseases or conditions can be both passive and active and is targeted toward "reportable diseases," meaning those diseases required by law to be reported to public health officials. Passive surveillance involves collecting, reviewing, and counting reports of suspected or diagnosed cases of disease. Active surveillance involves routinely performing systematic surveys or other investigations to actively detect cases that may not otherwise be reported.

The field of public health is highly varied and encompasses many academic disciplines, but the core areas have traditionally included the following, roughly ranked by order of importance:

1. Epidemiology/communicable diseases
2. Health education and communication
3. Preventive health
4. Public health laboratory
5. Environmental health
6. Health statistics

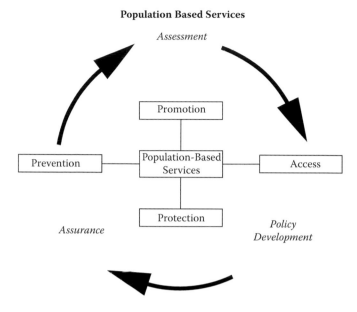

Population Based Services

Figure 1.18 Three main aspects of public health.

Source: Adapted from the Mississippi Department of Health

7. Healthcare licensure
8. Maternal and child health
9. Emergency medical services
10. Health services administration or local health administration (includes public health policy and economics)
11. Nutrition
12. Public health research

The Ten Essential Public Health Services

In 1994, the U.S. Public Health Service assembled and tasked a Public Health Functions Steering Committee to try to develop a working definition of public health and a framework for the responsibilities of local public health systems.[61] The resulting 10 essential public health services were issued (Table 1.2). I have assigned many of these 10 essential services to the older, traditional core areas, demonstrating that the new 10 public health services are not really new, just restatements or refinements of existing core functions.

Table 1.2 U.S. Public Health Service 10 Essential Public Health Services

Ten essential public health services (1994)	Traditional core areas of public health (1800s onward)
1. Monitor health status to identify and solve community health problems.	Health statistics
2. Diagnose and investigate health problems and health hazards in the community.	Epidemiology/communicable diseases
3. Inform, educate, and empower people about health issues.	Health education and communication
4. Mobilize community partnerships and action to identify and solve health problems.	–
5. Develop policies and plans that support individual and community health efforts.	Local health administration
6. Enforce laws and regulations that protect health and ensure safety.	Licensure or environmental health/communicable disease
7. Link people to needed personal health services and ensure the provision of healthcare when otherwise unavailable.	Health services administration or emergency medical services
8. Ensure a competent public and personal healthcare workforce.	–
9. Evaluate effectiveness, accessibility, and quality of personal and population-based health services.	Health statistics
10. Research for new insights and innovative solutions to health problems.	Public health research

Public Health Administration

The efficiency of public health organizations and their administration varies in different states depending on a variety of factors, such as degree of public and legislative support and financial appropriations available. Financial support for state health departments comes from the federal government and each state legislature. One source of federal funding is the CDC Preventive Health and Health Services (PHHS) block grants. Block grants give grantees the flexibility to prioritize the use of funds to fill funding gaps in programs that deal with leading causes of death and disability, as well as the ability to respond rapidly to emerging health issues, including outbreaks of foodborne infections and waterborne diseases. The PHHS block grant program provides all 50 states, the District of Columbia, 2 American Indian tribes, and 8 U.S. territories with funding to address their own unique public health needs and challenges in innovative and locally defined ways. Other federal funding opportunities have arisen through time and continue to do so. For example, since 2000, with the advent of West Nile virus (WNV) in the United States, the Department of Health and Human Services and the CDC provided approximately $180 million to state or local health departments to develop or enhance epidemiologic and laboratory capacity for WNV and other vector-borne diseases. This funding allowed for hiring of entomologists, surveillance programs for arboviruses, and laboratory testing capability for those viruses.

Enabling legislation of most states allows them to maximize purchasing power, minimize duplication, and maximize their administrative efficiency. This efficiency may vary somewhat in various states and territories because of differences in basic public health infrastructure and chain of command. For example, some states have one centralized public health system with nonoverlapping administration, purchasing, human resource management, and so on, whereas others may have multiple, independent city and county health departments.

States possess broad authority in matters pertaining to public health under "common law" and their constitutional police powers to protect the public (see Chapter 5). As far back as the late 1700s there was development of the idea of "medical police,"[20] and in 1807, New York City established a board of health and city inspector (Figure 1.19). By the mid-1800s health officials and even politicians in the United States realized the need for health police or health wardens to inspect, regulate, suppress, and prevent disease-producing circumstances. During this time the term *sanitary inspector* came into use.[20] In Great Britain, during the mid-1800s several far-reaching public health laws, such as the Sanitary Act,[2] were passed to protect public health and establish the role of *inspectors of nuisances.*

In the United States, creation and administration of health laws in the different states are generally accomplished through state boards or departments of health. In the former, authority is centered in the board and issues may be slower to resolve in that system. In the latter, the head of the department (state health officer or health commissioner) possesses executive authority, and prompt and decisive action is therefore possible. During a declared public health emergency, the health officer can do many things without legislative approval and skip ordinary checks and balances, such as bid processes.[62] Just what public health emergencies entail varies state by state, but generally refers to an occurrence or imminent threat of widespread damage, injury, or loss of life resulting from a natural phenomenon or human act.[62] In addition to state health officers with executive powers, there can be city, county, or district health officers as well. These health officers are usually the executive officers of the local city or county board of health, whose duty it is to enforce existing state or federal regulations. They may have been recruited from the ranks of local practicing physicians and given only part-time service. However, public

Figure 1.19 Board of Health doctor in a New York tenement.

Source: Photo courtesy National Library of Medicine

health work is now becoming more and more a distinct field, requiring specialized training beyond the MD degree, such as a master's in public health degree.

City or county health units are common in the United States, so the core public health function called local health administration may actually be one of the most important in a state health department since it relates to overseeing and guiding local health units. Routine fieldwork previously done in a county by staff from the state agency can be relegated to the locally supported county unit, which can, if required, obtain special assistance from subject matter experts in the various state divisions. Few states can provide a state organizational plan that can get as close to the people or function as efficiently in the field as the county health department. In

Figure 1.20 Public health service nurse visiting a rural family.

Source: Photo courtesy National Library of Medicine

fact, county health department doctors, nurses, and sanitarians have historically been on the cutting edge of delivering healthcare (Figure 1.20). City or county health departments may also be involved in the following activities:

- Health education
- Immunizations
- Food establishment permitting (retail food)
- Control of acute communicable diseases (disease outbreak investigators)
- Control of chronic communicable diseases (tuberculosis and venereal diseases)
- Control or investigation of special diseases (West Nile virus, etc.)
- Sanitation activities (safe public water and milk supplies and sanitary methods of sewage disposal)
- Child care and hygiene, including prenatal, infant, preschool, and school ages
- Maintenance of adequate records and preparation of reports

USE OF QUARANTINE IN PUBLIC HEALTH

Quarantine, derived from the Italian word "quaranta" (meaning 40), is a practice widely used for disease control since the 14th century.[1,2] Basically, it means separating people who have either contracted an infectious disease, or have been exposed to one, along with their goods and possessions, from other non-infected people for a period of 40 days. For millennia, the only effective way for humans to escape infection with various plagues was to avoid contact with infected people and their possessions. The Old Testament (OT) Mosaic Code included strict health-related rules for the people, such as washing hands, avoiding contact with dead people or animals, and isolation of sick people from the rest of the group. For example, in the Book of Leviticus, there are clear instructions for cases of leprosy. Lepers were to be isolated from others, their clothing burned, and their houses dis-infected or completely destroyed.[3] Although these OT health mandates were likely written down in their present form during the seventh century BCE, many scholars agree these concepts may have been practiced by the Hebrews since approximately 1500 BCE (3,500 years ago). Nobody knows for sure why a 40-day quarantine was ultimately selected as the time of isolation to prevent disease transmission. Some think it is related to Hippocrates' theories about acute illnesses, while others think it relates to the use of the number 40 in the Bible—40 years in the wilderness, 40 days in the desert, and so on. Nonetheless, quarantine is an effective and valuable way to slow or stop the spread of communicable diseases and is still in use today, although now modified to suit the particular disease ecology, incubation period, and period of infectivity for each disease. For example, quarantine was used extensively during the covid-19 pandemic during 2020, but was usually limited to 15 days.[4]

[1] **Matovinovac J**. *A short history of quarantine. Univ Mich Med Cent J. 1969;35:*224–228.

[2] **Tognotti E**. *Lessons from the history of quarantine, from plague to influenza A. Emerg Infect Dis. 2013;19:*254–259.

[3] **Gardner EJ**. *History of Biology. 3rd ed. Minneapolis:* Burgess Publishing Company; 1972.

[4] **CDC**. *When to quarantine if exposed to COVID-19. In:* CDC, www.cdc.gov/coronavirus/2019-ncov/if-you-are-sick/quarantine.html, accessed November 12, 2020; 2020.

Health Laws and Regulations

States do not have specific laws for every aspect of human health, safety, and welfare. Instead, there is only broad enabling legislation that allows the state health department to enact rules and regulations addressing myriad health-related issues (see Chapter 5). In most cases, rules and regulations are developed, then opened for public comment and input before being enacted by the department of health or, in some cases, by an oversight group such as a board of health. Drafts of proposed health legislation are usually carefully scanned by competent lawyers before enactment. Only laws capable of enforcement and with sound constitutionality can withstand legal challenges. Further, they must not exceed the scope of authority delegated by state legislatures. The ability to successfully enforce health statutes, particularly in cases that require jury trials, depends on existence of public sentiment (even agreement) and understanding of health-related laws.

Prosecution Issues

The practice of public health works because it has scientific evidence backing regulations and enforcement powers. Granted that health officials possess unique statutory authority to initiate legal proceedings against violators of public health laws, their success will in many instances depend on how intelligently they exercise this authority. Accordingly, tact and diplomacy are important skills for any health official, including public health entomologists. To avoid embarrassment and possible failure during enforcement actions, wise health officials recognize their power and authority and use them discerningly to address the public health challenges (see Chapter 5), making sure they have tried every reasonable avenue to resolve conflict and (then) have ample evidence supporting their case before going to court.

Notes

1. Harvey Shultz was an Army officer entomologist and Navy civilian entomologist.
2. The great sanitary act was passed in Britain in 1866, through the efforts of Dr. John Simon, and granted power to health authorities to carry out sanitary control measures.

References

1. Griscom JH. Discourse in 1855 before the New York Academy of Medicine. In: Rosenberg CE, ed. *Origins of Public Health in America*. New York: Arno Press; 1972:31–58.

2. USSC. Reports of the Sanitary Commission. In: The Sanitary Commission Bulletin, New York. 1863;1(4):1–200, December 15th issue.
3. Doane RW. *Insects and Disease*. New York: Henry Holt and Co.; 1910.
4. Pierce WD. *Sanitary Entomology: The Entomology of Disease, Hygiene, and Sanitation*. Boston: Richard G. Badger Company (The Gorham Press); 1921.
5. Clark HG. Superiority of sanitary measures over quarantine (lecture given in 1852). In: Rosenberg CE, ed. *Origins of Public Health in America*. New York: Arno Press; 1972:1–40.
6. Cushing E. *History of Entomology in World War II*. Washington, DC: Smithsonian Institution; 1957.
7. Marr JS, Malloy C. The ten plagues of Egypt (video). In: Cafe productions, in association with Oxford Scientific Films/BBC, London, UK; 1998.
8. Mumcuoglu YK, Zias J. Head lice from hair combs excavated in Israel and dated from the first century BC to the eighth century AD. *J Med Entomol.* 1988;25:545–547.
9. Hoeppli R. Parasitic diseases in Africa and the Western Hemisphere: early documentation and transmission by the slave trade. *Acta Tropica Suppl.* 1969;10:33–46.
10. Maco V, Tantalean M, Gotuzzo E. Evidence of tungiasis in pre-Hispanic America. *Emerg Infect Dis.* 2011;17:855–862.
11. Howard LO. A fifty year sketch history of medical entomology and its relation to public health. In: Ravenel MP, ed. *A Half Century of Public Health*. New York: American Public Health Association; 1921:412–438.
12. Ross R. On some peculiar pigmented cells found in two mosquitoes fed on malarial blood. *Br Med J.* 1897;18:1786–1788.
13. Reed W, Carroll J, Agramonte A. Experimental yellow fever. *Trans Assoc Am Phys.* 1901;16:45–71.
14. Reed W, Carroll J, Agramonte A, Lazear JW. The etiology of yellow fever—a preliminary note. In: Proceedings of the 28th Annual Meeting of the American Public Health Association. 1900;26:37–53.
15. Nicolle C, Comte C, Conseil E. Experimental transmission of the exanthematic typhus by the body louse. *C R Acad Sci.* 1909;149:486–489.
16. Pierce WD. The deadly triangle: A brief history of medical and sanitary entomology. In: Natural History Museum of Los Angeles County, Los Angeles, CA. 1974:138 pp.
17. Snyder JC. The philosophy of disease control and the population explosion. In: WHO/PAHO, ed. *The Control of Lice and Louse-Borne Diseases*. Washington, DC: World Health Organization, Scientific Publication Number 263; 1972:7–13.
18. Peterson RKD. Insects, disease, and military history. *Am Entomol.* 1995;Fall:147–160.
19. Crosby MC. *The American Plague*. New York: Berkley Books; 2006.
20. Freedman B. *Sanitarian's Handbook: Theory and Administrative Practice for Environmental Health*. 4th ed. New Orleans: Peerless Publishing Co.; 1977.
21. Mattingly PF. Origins and evolution of the human malarias: the role of the vector. *Parassitologia.* 1973;15:169–172.
22. Bruce-Chwatt LJ. *Essential Malariology*. 2nd ed. New York: John Wiley and Sons; 1985.
23. Rogers K. Malaria through history. In: Encyclopedia Britannica online; 2020. www.britannica.com/science/malaria/malaria-through-history.

24. de Meillon B. The control of malaria in South Africa by measures directed against the adult mosquitoes in habitations. *Q Bull Health Organ League Nat.* 1936;5:134–137.
25. Ackerknecht EH. Malaria in the United States. In: CIBA Symposia, Summit, New Jersey. 1945;7(3–4):63–68.
26. Faust EC. The distribution of malaria in North America, Mexico, Central America, and the West Indes. In: Moulton FR, ed. *Human Malaria with Special Reference to North America and the Caribbean Region.* Washington, DC: AAAS; 1941:8–18.
27. Faust EC. Malaria mortality and morbidity in the United States for the year 1941. *J Nat Malaria Soc.* 1942;2(1):39–43.
28. Bass CC. Studies on malaria control. I. The relative frequency of malaria in different ages and age groups in a large area of great prevalence. *South Med J.* 1919;12:456–460.
29. Bass CC. The passing of malaria. *New Orleans Med Surg J.* 1927;79:713–719.
30. Levine RS, Peterson AT, Benedict MQ. Distribution of members of *Anopheles quadrimaculatus* say s.l. (Diptera: Culicidae) and implications for their roles in malaria transmission in the United States. *J Med Entomol.* 2004;41(4):607–613.
31. Wilkerson RC, Reinert JF, Li C. Ribosomal DNA ITS2 sequences differentiate six species in the *Anopheles crucians* Complex. *J Med Entomol.* 2004;41:392–401.
32. Perez M. An anopheline survey of the State of Mississippi. *Am J Hyg.* 1930;11:696–710.
33. Barber MA, Komp WH. Breeding places of *Anopheles* in the Yazoo-Mississippi Delta. *Public Health Rep.* 1929;44:2457–2462.
34. Hoffman FL. Malaria in Mississippi and adjacent states. *South Med J.* 1932;25:657–662.
35. von Ezdorf RH. Malarial fevers: prevalence and geographic distribution in Mississippi, 1913. *Public Health Rep.* 1914;29:1289–1300.
36. Harden FW, Hepburn HR, Ethridge BJ. A history of mosquitoes and mosquito-borne diseases in Mississippi 1699–1965. *Mosq News.* 1967;27(1):60–66.
37. Langmuir AD. Evolution of the concept of surveillance in the United States. *Proc R Soc Med.* 1971;64:9–12.
38. Busbey LW. *Uncle Joe Cannon.* New York: Henry Holt and Company; 1927.
39. Faust EC. Malaria incidence in North America. In: Boyd MF, ed. *Malariology: A Comprehensive Survey of All Aspects of this Group of Diseaes from a Global Standpoint.* Philadelphia: W.B. Saunders Company; 1949:749–763.
40. Watson RB, Maher HC. An evaluation of mosquito-proofing for malaria control based on one year's observations. *Am J Epidemiol.* 1941;34:86–94.
41. Dowling O. Malarial infection in the lower Mississippi Valley. In: American Society of Tropical Medicine, Proceedings from the Annual Meeting, Chicago, IL. 1924;5(S1–4):461–468.
42. Brierly WB. Malaria and socio-economic conditions in Mississippi. *Social Forces.* 1945;23:451–459.
43. Hinman A. 1889 to 1989: a century of health and disease. *Public Health Rep.* 1990;105:374–380.
44. Underwood FJ. The control of malaria as a part of full-time county health department program. *South Med J.* 1929;22:357–359.
45. Bradley GH, Bellamy RE, Bracken TT. The work of state board of health entomologists on malaria control. *South Med J.* 1940;33:892–894.

46. Mountin JW. A program for the eradication of malaria from continental United States. *J Nat Malaria Soc.* 1944;1:61–73.
47. Humphreys M. *Malaria: Poverty, Race, and Public Health in the United States.* Baltimore, MD: Johns Hopkins University Press; 2001.
48. Barrier AK. Screening in Sharkey County, Mississippi. *South Med J.* 1932;25:662–664.
49. Barber MA. The history of malaria in the United States. *Public Health Rep.* 1929;44:1–13.
50. Faust EC. The history of malaria in the United States. *Am Sci.* 1951;39:121–130.
51. Shultz HA. 100 years of entomology in the Department of Defense. In: Adams J, ed. *Insect Potpourri: Adventures in Entomology.* Gainesville, FL: Sandhill Crane Press; 1992:61–72.
52. Streck-Havill T. Dr. Clara Ludlow: from music to mosquitoes. In: The Micrograph, U.S. Museum of Health and Medicine; 2019. www.medicalmuseum.mil/micrograph/index.cfm/posts/2019/dr_clara_ludlow_from_music_to_mosquitoes.
53. Kahn EJ, Jr. *G.I. Jungle: An American Soldier in Australia and New Guinea.* New York: Simon and Schuster; 1943.
54. Navy. Military medicine in Operations Desert Shield and Desert Storm. In: The Navy Forward Laboratory, Biological Warfare Detection, and Preventive Medicine; 1998. https://gulflink.health.mil/pm/pm_refs/n38en014/med_navy.htm.
55. Hyams KC, Hanson K, Wignall FS, Escamilla J, Burans J, Woody JN. The impact of infectious diseases on the health of U.S. troops deployed to the Persian Gulf during Operations Desert Shield and Desert Storm. *Clin Infect Dis.* 1995;20:1497–1504.
56. Cope SE, Schoeler GB, Beavers GM. Medical entomology in the United States Department of Defense: challenging and rewarding. *Outlooks Pest Manag.* 2011;22:129–133.
57. Snetsinger R. *The Ratcatcher's Child: History of the Pest Control Industry.* Cleveland, OH: Franzak and Foster Co.; 1983.
58. Mattern S. The art of medicine: Galen and his patients. *Lancet.* 2011;378:478–479.
59. Boyd MF. *Preventive Medicine.* Philadelphia: W.B. Saunders Co.; 1932.
60. Frieden TR. The future of public health. *N Eng J Med.* 2015;373:1748–1754.
61. CDC. The ten essential public health services. In: Centers for Disease Control and Prevention, Environmental Health Services, Atlanta, GA; 1994. www.cdc.gov/nceh/ehs/Home/HealthService.htm (accessed August 23, 2010).
62. Haffajee R, Parmet WE, Mello MM. What is a public health emergency? *N Engl J Med.* 2014;371:986–988.

chapter two

Pest Control in Modern Public Health

Role of Pesticides in Public Health

The public health entomologist is the main resource person in a health department concerning pests, pest control, and pesticides. Accordingly, he or she should be knowledgeable about the subject. Health officials generally recognize that pesticides are important public health tools. Technically, they are defined as chemical substances intended to prevent, destroy, or repel pests. The term *pesticide* is usually further subdivided into more specific terms, such as fungicide (kills fungus), herbicide (kills plants), acaricide (kills mites and ticks), avicide (kills birds), and so on. Pesticides may be either synthetic chemicals or naturally occurring substances such as certain inorganic dusts, bacterial toxins, or plant derivatives.

Pesticide Laws and Registration

The sale and use of pesticides is strictly regulated by federal and state laws. In the United States, the earliest law that included language about pesticides was the Pure Food Law of 1906, which has been amended several times into what is now commonly known as the Federal Food, Drug, and Cosmetic Act. This law contains provisions to protect the consumer from pesticide-contaminated foods, among other things. The Federal Insecticide Law was enacted in 1910, and was later replaced by a much stronger version known as the Federal Insecticide, Fungicide, and Rodenticide Act (FIFRA) in 1947. FIFRA also has been amended several times, but is still the primary legislation concerning manufacture, registration, labeling, and use of pesticides in this country.

The U.S. Environmental Protection Agency's (EPA) Office of Pesticide Programs (OPP) is tasked with registering or licensing pesticides for use in the United States and, along with its 10 regional offices and other EPA programs, works on a wide range of pesticide issues and topics. All pesticide uses and applications, whether chemical or biological pesticides, are required under FIFRA to undergo thorough analysis by the EPA to determine whether they pose adverse risks to the environment or human health. Each pesticide application is subject to rigorous scientific analysis followed by a risk management (also known as risk-benefit) determination. Ultimately, each pesticide registration is either denied or granted with or without restrictions.

FACTORS THE EPA CONSIDERS IN REGISTERING PESTICIDES

- Product chemistry and inherent toxicity
- Product efficacy
- Any public health benefits
- Any economic benefits
- Potential negative health effects (especially in children)
- Potential negative effects on nontarget animals and the environment

Manufacturers of products needing registration must submit (for each product) ecological effects data for nontarget animals to OPP, and in the case of a public health pesticide (if claimed on the label), must submit efficacy studies for the target pests (i.e., mosquitoes) to support claims of efficacy. Industry applicants must also submit data on health effects, product chemistry, and economic benefits. OPP's ecological and environmental effect teams, for both conventional pesticides and biopesticides, assess the risk of pesticide exposure and potential effects on nontarget wildlife, plants, soil, and water (ground or surface). The health effects teams assess risk from exposure and negative effects in humans; specific groups of the population that are at greater risk of exposure and poisoning (i.e., children) are considered to determine a minimum acceptable (threshold) exposure risk. OPP's economic analysis team assesses associated benefits from the registration of each pesticide. Finally, EPA risk managers consider risk assessments and benefits analysis and reach a risk-benefit determination. This sometimes includes restrictive language required to be placed on the label (i.e., reducing application rates, altering allowed timing of applications, etc.) or excluding certain proposed uses or limiting the pesticide use to certain geographic areas. For public health pesticides specifically, registrants are prohibited from using language claiming to control or mitigate microorganisms in a way that links the microorganism to a threat in human health (including but not limited to disease-transmitting bacteria or viruses). For example, the label may not say "controls ticks that carry Rocky Mountain spotted fever" or "controls mosquitoes that can transmit malaria or encephalitis," but must simply say "controls ticks" or "controls mosquitoes."

Obtaining Pesticide User Certification and Licenses

There are different types of licenses, permits, and certificates required for individuals who recommend, apply, or sell pesticides. Written examinations required for commercial licenses and certificates are administered by a variety of state agencies, but often they reside in each state's Department of Agriculture (Table 2.1). While every state follows the same federal regulations for pesticide use, program details are different in every state. For example, some states have more certification/licensing categories than other states, and the length of certification varies between states, and renewal methods vary.

Pesticide Applicator Certifications

Under provisions of various state pesticide laws, a state regulatory agency carries out various activities to certify commercial and private applicators of pesticides, cooperate with the EPA on enforcement of federal pesticide laws, inspect records of applications of restricted-use pesticides, and investigate pesticide misuse complaints.

Private Applicator Certification. Private applicators are generally farmers. A private applicator is a certified applicator who uses or supervises the use of restricted-use pesticides to produce an agricultural commodity on his or her own land, leased land, or rented land, or on the lands of his or her employer. Private applicators must be at least 18 years old. In most states, County Extension agents provide training for private

Table 2.1 Examples of State Agencies That Regulate Pesticide
Certification and Licensing.

State	Regulatory Agency
California	Dept. Pesticide Regulation
Indiana	Office of the State Chemist
Florida	Dept. Agriculture and Consumer Services
Mississippi	Dept. Agriculture and Commerce
Montana	Dept. Agriculture
New York	Dept. Environmental Conservation
Oregon	Dept. Agriculture
South Dakota	Dept. Agriculture
Texas	Dept. Agriculture

pesticide applicators and assure completion of the application process for their certification. The agents schedule private applicator training sessions and conduct them in their counties according to policies handed down by the appropriate pesticide regulatory agency. Private applicator certification expires after 3–5 years. To renew their certification, private applicators must attend a training session and pass a written examination during the year prior to their expiration date.

Commercial Applicator Certification. The sale and application of certain types of pesticides are restricted by law because of the probability of adverse effects on humans and the environment if these pesticides are improperly used. Pesticides classified as "restricted-use" may be applied only by, or under supervision of, certified applicators. These applicators must attend training sessions and pass a written examination before they receive certification. Certification categories vary widely by country, state, or province, but examples are provided in Table 2.2.

Certified Pesticide Applicator Renewals. A certified pesticide applicator certification is not a license, and does not allow the person to advertise, solicit business, or perform work for a fee as a pesticide applicator.

Table 2.2 Some Examples of Pesticide Certification Categories.

- Agricultural Plant
- Agricultural Animal
- Forest
- Ornamental
- Turf Management
- Seed Treatment
- Aquatic
- Right-of-Way
- Industrial, Institutional, Structural, and Health-Related
- Public Health
- School IPM
- Commercial Wildlife Management
- Lawn and Ornamental
- Landscape Maintenance
- Regulatory
- Demonstration and Research
- Aerial
- Wood Preservation and Wood Products Treatment
- Fumigation
- Field Fumigation
- Microbial Pests

That would require a license (see the next section). Commercial pesticide applicator certification is valid for a few years (depends upon state or province). Certified commercial applicator certification is renewable by attending approved recertification training within a prescribed time period prior to expiration, or by reexamination. Generally, recertification training takes about a half day and qualifies for renewal of certificates in each of the categories. Often, no examination is required for recertification. After appropriate online or in-person training, the certificate will be reissued.

Pest Control Licenses

A professional license is required for persons advertising and/or soliciting business to control pests or weeds, or to perform various activities such as landscape gardening, tree surgery, or soil classification. Control-type licenses are offered in many categories (Table 2.3). These types of

Table 2.3 Some Examples of Both Pest and Weed Control License Categories.

- Agricultural Pest Control
- Forest Pest Control
- General Pest and Rodent Control
- Household, Structural, and Industrial
- Medical, Veterinary, and Public Health
- Orchard and Nut Tree
- Ornamental
- Public Health
- Mosquito and Biting Fly
- Wood Destroying Insect Control
- Horticultural Pest Control
- Domestic Animal Pest Control
- Government Applicators
- Public Utility Applicators
- Right-of-Way Weed Control
- Agricultural Weed Control
- Aquatic Weed Control
- Tree Surgeons
- Soil Classification
- Fumigation
- Aerial Application
- Regulatory Pest Control
- Pesticide Dealer
- Demonstration and Research

licenses generally expire 3–5 years from date issued. To renew a control-type license, holders must submit a form prescribed by their state regulatory agency and either attend a training course approved by that agency within the past 12 months or pass an examination. A pest control license is required for persons charging fees for applying pesticides and/or making recommendations for control of pests.

Pest Management Consultant Licenses

A pest management consultant license is required for persons making recommendations and charging fees for entomology, plant pathology, or weed control services. A consulting license does not allow for actual spraying or control of these pests. This license is offered in three major areas: entomology, plant pathology, and weed control, and there are various categories within each area:

> *Entomology*: Persons charging fees for consultation, advice, or recommendations in the control of insects and/or vertebrate pests must be licensed in the appropriate entomology category listed:
> * Agricultural
> * Forest
> * Household, Structural, and Industrial
> * Medical, Veterinary, and Public Health
> * Orchard and Nut Tree
> * Ornamental
>
> *Plant Pathology*: Persons charging fees for consultation, advice, or recommendations in the control of plant diseases must be licensed in the appropriate plant pathology category listed:
> * Agricultural Plant
> * Forest Plant
> * Orchard and Nut Tree Plant
> * Ornamental and Shade Tree Plant
>
> *Weed Control*: Persons charging fees for consultation, advice, or making recommendations in the control of weeds must be licensed in the appropriate weed control category listed:
> * Agricultural
> * Aquatic
> * Forest and Right-of-Way
> * Ornamental and Turf
> * Industrial or Commercial

Permits

Permits generally pertain only to pest and weed control firms. Each branch location for a pest control and/or weed control license-holder must have at least one licensee or one permit holder (sometimes also called an "operator") for each category under which the licensee is soliciting and/or performing work. Most states/provinces require that each branch office have a license holder or permit holder directly supervising[1] the daily activities of the office. An employee who doesn't meet the qualifications to become licensed may apply to take permit exam(s) for the area(s) of work performed from that branch location. To qualify for permit examinations one must be a *bona fide* employee of the license holder and have a valid state registered technician identification card. Permit holders are not allowed to own and operate a company (that requires a license).

NPDES Guidelines Impacting Pesticide Applications

In the United States, discharge of pollutants into waterways is regulated under the Clean Water Act of 1972 (CWA) through its National Pollutants Discharge Elimination System (NPDES). In the early 2000s, citizens and environmental groups began to insist that widespread pesticide use— and specifically mosquito spraying in areas over 6,400 acres/year—fall under the Clean Water Act, requiring additional regulatory oversight and (thus) NPDES permits.[1] Pesticides applied ultra-low volume (ULV) for mosquito control have a tendency to drift and may deposit onto water (Figure 2.1).

Figure 2.1 Pesticides used for mosquito control may drift into bodies of water.

Source: Photo copyright 2020 by Jerome Goddard, Ph.D.

Environmental groups argued in court that deposition of these pesticides in water is a violation of the CWA, and further, that this is tantamount to discharging a pollutant into the navigable waters of the United States without a permit.[1] A court ruling in 2009, commencing October 31, 2011, agreed with this assertion, requiring NPDES permits for the application of mosquito control pesticides (both biological and synthetic), whenever an application results in a pesticide residue, however minimal, entering waters of the United States. These permits, at least from the environmental groups' perspective, help monitor chemicals entering the nation's waterways, help mitigate downstream adverse effects, and keep drinking water safe. The 2009 court ruling negated an earlier policy issued by the EPA in 2006 clarifying circumstances in which a Clean Water Act NPDES permit is not required for discharges from the application of pesticides to or around water. Subsequently, the American Mosquito Control Association (AMCA) got involved, working with the EPA and authorized states to craft NPDES pesticide "general" permits that minimized potential impacts on mosquito control and its public health importance. Those general permits went into effect in six states where the EPA is the permitting agency. All other U.S. states have subsequently drafted their own permits for pesticide discharge. From the viewpoint of the AMCA, all NPDES permits still result in scarce public funds being spent on duplicative regulatory requirements, administrative fees, and legal costs. This is because pesticides are already effectively regulated for uses in and near water under the registration process required by FIFRA. Under FIFRA, the EPA requires an extensive range of scientific studies, which determine potential impacts on water quality and aquatic species. If needed, the EPA has broad powers to require additional information, and does so where necessary to ensure that it thoroughly understands a pesticide's risks to people and the environment.

The AMCA highlights that the need for mosquito control personnel to utilize their time and resources efficiently are more important than ever, especially with the emergence of Chikungunya and Zika viruses affecting the U.S. population. The United States has also seen significant impacts of other diseases such as West Nile virus, Eastern equine encephalitis, and Dengue in the past decade and the potential for new and emerging viruses is ongoing. It is important that funds are spent on the actual control of mosquitoes and not duplicative regulations. In 2019, a bill was introduced in the U.S. House of Representatives, the REDTAPE Act (H.R. 890), to establish exemptions to the NPDES/mosquito control issue. The bill modifies requirements governing the use of pesticides in or near navigable waters, and specifically, prohibits the EPA or states from requiring NPDES permits for discharges of pesticides into navigable waters if the pesticides are (1) registered, (2) used for their intended purposes, and (3) used in compliance with their pesticide label requirements. However, as of this writing, this Act has not been approved by the U.S. Congress nor signed into law by the president. Therefore, all public health, municipal, and community mosquito control programs should check with their local regulatory agency for up-to-date guidance about NPDES or other permits for widespread mosquito spraying.

History and Current Status of Pesticides

From ancient times, humans have occasionally used chemical compounds to try to ward off pests, for example, sulfur for itch mites[2] (we now know that sulfur is an effective acaricide that is relatively safe for human use).[3] However, during the late 1800s copper sulfate, nicotine, fluorides, pyrethrum powder, and arsenicals came into widespread use as insecticides (Figure 2.2). For example, sodium silicoflouride was used to control ectoparasites on livestock as well as crawling insects in houses and buildings (Figure 2.2D). These inorganic compounds, while not miracle drugs, were

Figure 2.2 Some examples of the first pesticides.

certainly better than nothing, but large dosages were required. In the 1930s, scientists found that the synthetic compound paradichlorobenzene, which had been used extensively for clothes moth control, was effective against peachtree borers. Then scientists discovered that substituted phenolic compounds had insecticidal properties, and thus began the systematic search for related synthetic compounds. DDT, one of the most famous of all pesticides, was first synthesized in 1874, but its insecticidal properties were not discovered until 1939[4] (Figure 2.3). DDT was used with great success in the second half of World War II to control malaria and typhus among civilians and troops, being sprayed directly on people and property (Figure 2.4). Other chlorinated hydrocarbons were soon developed, such as lindane, endrin, aldrin, chlordane, and many others. These compounds had very long residual effects, still killing insects 10–30 years after application.

Pesticide development and use can be roughly grouped into several phases. As mentioned, the first-generation pesticides consisted of such things as kerosene (and other oils), sulfur, arsenic, and nicotine sulfate (Figure 2.5). These were often applied by Flit guns (named after an insecticide used for fly control) or other primitive dusting or application equipment (Figure 2.6). Mosquito control personnel used to "oil" storm drains to kill developing mosquito larvae (Figure 2.7). Also, during the early 1900s pyrethrins, derived from dried flower heads of Chrysanthemum plants, were used as pesticides. Interestingly, they are still a mainstay in pest control as they are considered "natural" or botanical insecticides

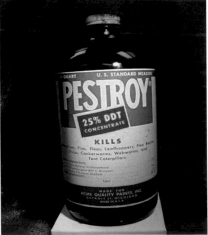

Figure 2.3 Some examples of organochlorine pesticides.

Figure 2.4 DDT being sprayed on people and property.

Source: From the United States Army and the United States Department of Agriculture

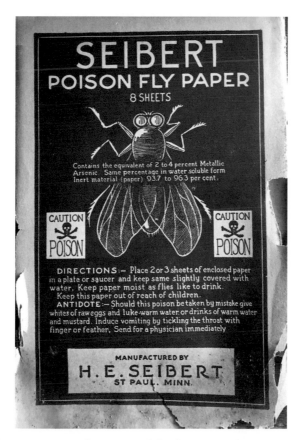

Figure 2.5 Arsenic was used in many of the first pesticides.

Source: Copyright 2021 by Jerome Goddard Ph.D.

Figure 2.6 Examples of historical pesticide application equipment.

Figure 2.7 Inspector oiling sewer catch basin for mosquito control.

Source: Photo from the New Jersey Agricultural Experiment Station, 1911

(Figure 2.8). During and immediately after World War II, far more potent second-generation pesticides were developed, including the chlorinated hydrocarbons such as DDT, and later in the 1950s to 1960s, several new carbamates and organophosphates were added to the repertoire of insecticide chemistries. These were used extensively until the early 1970s, when environmental and ecological concerns were raised by Rachel Carson in her famous book, *Silent Spring*, and that is when many of the chlorinated

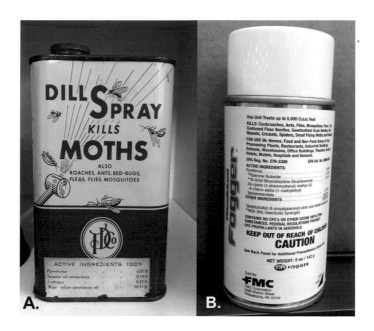

Figure 2.8 Pyrethrins have been used as effective pesticides since the early 1900's.

Source: Photo copyright 2021 by Jerome Goddard, Ph.D.

hydrocarbons such as DDT were banned. These environmental concerns also led to the eventual restriction of many of the carbamates and organophosphates, leading the way for research and development of third-generation products such as pheromones, insect hormones, host attractants, and other biocontrol agents. More recently, the green movement has spurred the use of even more "safe" repellents and pesticides, such as those composed of food-grade essential oils (Figure 2.9). As for traditional, residual pesticides, they are certainly still in existence (mostly pyrethroids and several newer classes of chemistry), but used sparingly or in conjunction with other nonchemical control methods, called integrated pest management (IPM) (Figure 2.10).

In the last two decades, pesticide use has increasingly been scrutinized in this country, with some segments of the population demanding elimination of most (if not all) uses of the substances. However, it is important to realize that—from the public health perspective—pesticides serve a valuable function in society primarily by preventing/controlling infectious diseases carried by insects and other arthropods.[5]

Infectious diseases are making a strong comeback after a lull in the years following World War II. The ability of disease germs to adapt to the human defense system and intense pressure from antibiotic use, combined

Figure 2.9 Some examples of natural or herbal pesticides and repellents.

Figure 2.10 Graphic showing components of integrated pest management.

Source: From GIE Media, used with permission

with changes in society, has contributed to this comeback. Also, there are now several "new" or emerging diseases, including covid-19, severe acute respiratory syndrome (SARS), Legionnaires' disease, Lyme disease, ehrlichiosis, toxic shock syndrome, and Ebola hemorrhagic fever. In just the last three decades we have seen the appearance of a new strain of bird influenza that attacks humans, a human form of "mad cow" disease, and new drug-resistant forms of *Staphylococcus aureus*. These new or emerging infectious diseases have raised considerable concern in the medical community about the possibility of widespread and possibly devastating disease epidemics (see Chapter 6 for more discussion of emerging diseases).

**FACTORS LEADING TO EMERGING
INFECTIOUS DISEASES**

- Human population increase
- Ecological and environmental changes/disruptions
- Genetic changes in pathogens (mutations)
- Pesticide resistance
- Drug resistance (to the medicines used to treat them)
- International travel

Lack of Pesticides: A Cause for Future Concern

Humans are now in a precarious situation. The entire ecosystem—including plant and animal life on earth—is being negatively affected by human civilization. People once lived in far-removed, relatively isolated groups. Now we are all essentially one large community. Further, things such as population increases, building cities in or near jungles, and widespread and frequent international air travel are providing opportunity for an outbreak of a great plague. For example, in 1950, a trip from London to Hong Kong required 5–6 days, and the number of passengers on internationally scheduled airlines in that year was approximately 6 million. By 1970, the London–Hong Kong trip had been reduced to 8–10 hours, and the number of passengers had increased to 70 million.[6] More recently, the number of international departures from U.S. airports doubled from 20 million to nearly 40 million between 1983 and 1995.[7] A person hiking in the Amazon jungles today might be in New York City (or your town) tomorrow. Should one or more new emerging vector-borne diseases begin to spread, control of the epidemic would be difficult. As an example of the quick spread of disease agents, consider swine flu, which was first discovered in Mexico in late March 2009, and by the first week of May (6 weeks later) had spread to many places worldwide, from New York to New Zealand. If an emerging disease agent is a virus, specific treatments are unavailable (or, at least, untested against most arboviruses). The only way to stop a viral vector-borne illness is to kill the vectors to a low enough level to interrupt virus transmission. If the vector is a flying insect, control of an epidemic is even harder. Compounding all of this, many insect species are resistant to many of the traditional insecticides used to control them.

GMO Mosquitoes and Other Emerging Control Technologies

In light of the afore-mentioned potential loss of insecticides as public health tools in the fight against mosquito-borne diseases, there is significant need to explore entirely new methods of mosquito control. One emerging method is the use of genetically modified organisms (GMO), specifically genetically modified mosquitoes (GMM), to either curb mosquito population sizes or restrict their vector competency. However, use of genetically modified mosquitoes to reduce incidence of disease is controversial. Currently, there are several approaches: (1) population replacement—developing strains of mosquitoes that are resistant to infection with disease agents and trying to replace wild populations of mosquitoes with these transgenic ones. Making this possible, there are CRISPR-Cas9 based gene-drive mechanisms to spread these genes throughout the population, allowing super-Mendelian inheritance of a transgene with the potential to modify large insect populations over a short time frame;[8–10] (2) release of millions of sterile mosquitoes into an area to mate with native female mosquitoes for population suppression (the most controversial way is by using *Wolbachia*-induced cytoplasmic incompatibility);[11] and (3) release of genetically modified male mosquitoes (such as *Aedes aegypti*) containing a dominant lethal gene that will kill their offspring after mating with wild mosquitoes in an area. This particular self-limiting technology is currently being used by Oxitec (see the next section). The fear with genetic approaches is that transgenes placed in GMMs could spread further, quicker (see the section on gene drives that follows), into non-intended hosts in the release sites as well as neighboring countries.[12] All these ethical, legal, and social issues have yet to be resolved.

Gene-editing technologies have expanded possibilities of vector control by allowing researchers to edit target genes, analyze gene functions, and even gene expression (up—or downregulation). There are at least 12 classes of these genetic technologies used on disease vectors, but here only the Oxitec GMM and CRISPR-Cas9 technologies are discussed.

CRISPR-Cas9 technology. CRISPR-Cas9 is an RNA-guided endonuclease genome-editing technology known for its versatility, low cost, specificity, and simplicity. After a single guide RNA (sgRNA) recognizes a target sequence, the Cas9 nuclease makes double-stranded cuts near a specific motif sequence. This allows for multiple DNA breaks at different sites *in vitro*,[13] and scientists can manipulate the system through alteration of sgRNA. CRISPR-Cas9 is constantly improving, either through its application in different mosquito vectors, through streamlined protocols, or improved methodologies of delivery into female mosquito germlines, and has even shown heritability of modified genomes through successful subsequent generations maintaining the genome edits.[8,14–19] For example, CRISPR-Cas9 has been used to show that ectopic expression of *Nix*, a male determining gene, in female mosquitoes results in male genital development.[20] This opens a path for future control strategies involving death of females, or female-to-male conversion, through genomic manipulation of these male-determining genes, an idea supported by research in *Drosophila* fruit flies.[21,22] In addition, the CRISPR-Cas9 system has helped identify olfactory genes responsible for host selection chemosensory processes in *Anopheles coluzzi*, a malaria vector.[23] Manipulation of olfactory systems in mosquitoes could lead to decreased ability to find a host, or even decreased anthropomorphic behaviors driving mosquito biting habits. Further, Zhu *et al.* discovered that a knockout of the *Methoprene-tolerant (Met)* gene coding for a juvenile hormone receptor led to a black larval phenotype in the fourth instar and prevented pupation.[24] Manipulation of this gene also could be beneficial in mosquito control.

Oxitec GMO mosquitoes. Oxitec is a UK-based biotechnology company focused on using genetically modified insects to control insect populations across the globe.[25] In partnership with the Bill and Melina Gates Foundation and a few agricultural companies, Oxitec researchers refined the Sterile Insect Technique (SIT) for mosquito control, particularly using *Aedes aegypti*. Oxitec researchers engineered a self-limiting system using tTAV, a tetracycline-repressible transcriptional activator.[26–28] The tTAV activator is controlled by presence of the antibiotic tetracycline, and when grown with tetracycline present, mosquitoes will not express tTAV, allowing for mosquito survival. However, in absence of tetracycline, tTAV is expressed, causing mosquito death due to toxicity.[28] This is a "kill switch" of sorts. Additionally, this technology is female-specific. Genetically modified males carrying tTAV mate with wild females in nature, and female offspring die while male offspring carrying this gene survive to transmit it to the next generation.[29]

This technology has already been implemented in nature. For example, experimental trials of OX5034 mosquitoes were conducted in São Paulo, Brazil in 2018–19[30] where male *Aedes aegypti* mosquitoes containing the tTAV gene were released and their abundance monitored. In those trials, Oxitec reported an average of 89% suppression over 4 weeks with a low release rate of mosquitoes, and an average of 93% suppression over 4 weeks using a high release rate of mosquitoes.[30]

On May 1, 2020, Oxitec received approval from the Environmental Protection Agency (EPA) to launch pilot projects for OX5034 male *Aedes aegypti* mosquitoes in the USA.[31,32] This came after several years of Oxitec's approval request(s) being bounced from the U.S. Department of Agriculture, to the U.S. Food and Drug Administration (FDA), to finally the EPA in 2018. Subsequently, in June of 2020 the state of Florida granted

Oxitec experimental use permission and in August of 2020, Florida Keys Mosquito Control District (FKMCD) Board of Commissioners granted approval for a pilot project to be conducted in the Florida Keys.[33,34] On April 29, 2021, the first boxes of mosquitoes were placed in an undisclosed location in the Florida Keys.[35]

Gene drives. So-called "gene drives" allow genetic elements to be inherited and spread through a population at a rate much faster than predicted by Mendelian genetics, and this is a rapidly emerging approach to vector and pathogen control which has been applied in mosquitoes.[8,9,36,37] Gene drives rely on the use of a driving endonuclease gene (DEG), which includes CRISPR-Cas9 systems (and a couple of other lesser known technologies). There are two ways gene drives can be applied in fieldwork: (1) suppression of insect populations through targeting genes associated with insect fecundity, lifespan, or mortality, or (2) reduction of vector competency through changing genes of interest associated with pathogen infection of the insect.[10,38,39]

Potential problems associated with gene drives and GMMs. As illustrated by the recent use of Oxitec mosquitoes for suppression of *Aedes aegypti* mosquitoes, genetically modified mosquitoes are an important emerging technology intended to minimize the public health burden mosquitoes impose; however, releasing these synthetically modified organisms into nature and letting them multiply and reproduce unabated raises concerns. In some cases, gene drives and the genome modifications proposed could lead to population collapse(s), either completely or nearly eliminating mosquito species from the environment of a region, raising ecological and ethical concerns.

One major concern of entomologists and ecologists is the possible fitness cost to the mosquito, which could lead to transgene silencing, the evolution of "resistance," or even a restoration of wildtype fitness.[37] In the latter, the wildtype mutant would quickly rebound. A rise in drive-resistant alleles in natural populations, especially after mating of transgenic mosquitoes and wildtype mosquitoes, could result in unimagined and undesirable downstream effects.[40] Increasing numbers of drive-resistant insects would eliminate the drive through natural selection since the drive itself would likely impart some level of fitness reduction.[41] Therefore, we must assess the evolutionary stability of gene drives in nature. Further, effects of releasing a genetically modified insect into the environment are still largely unknown. There could be potential downstream off-target effects such as a potential risk of other mutated pests. Detecting these mutated off-target pests would prove near impossible once GMMs are unleashed in the environment.[42] A successful gene drive would require mating between mutants and wildtype mosquitoes, so reliability of the gene to be heritable and effective in these crosses after multiple generations needs to be accurately assessed.[41,43]

Additionally, genomic editing can be very expensive, especially on the scale of altering entire populations of insects, not just lab colonies. Yes, genome editing systems are becoming more accessible and simpler to use, but applying this technology to a system that will be released to the wild requires (first of all) certainty that the gene drive is working correctly. Genomic editing, and especially those technologies involving cutting the DNA, requires precision of the cut and some means of repairing the cut sequences using homologous recombination instead of non-homologous recombination. These molecular conditions determine whether a gene drive will work at all.

Note

1. Exactly what constitutes "direct supervision" varies widely by regulatory agency.

References

1. Leintz RE. Is FIFRA enough regulation? Failure to obtain a NPDES permit for pesticide applications may violate the Clean Water Act. *Chicago-Kent L Rev*. 2004;79:317–341.
2. Cushing E. *History of Entomology in World War II*. Washington, DC: Smithsonian Institution; 1957.
3. Yu SJ. *The Toxicology and Biochemistry of Insecticides*. Boca Raton, FL: CRC Press; 2008.
4. USDA. DDT and other insecticides and repellents developed for the armed forces. In: USDA, Misc. Publ. No. 606; 1946:71 pp.
5. Rose RI. Pesticides and public health: integrated methods of mosquito management. *Emerg Infect Dis*. 2001;7:17–23.
6. Bruce-Chwatt LJ. Global problems of imported disease. *Adv Parasitol*. 1973;11:75–114.
7. Gubler DJ. Arboviruses as imported disease agents: the need for increased awareness. *Arch Virol*. 1996;11:21–32.
8. Gantz VM, Jasinskiene N, Tatarenkova O, et al. Highly efficient Cas9-mediated gene drive for population modification of the malaria vector mosquito *Anopheles stephensi*. *PNAS*. 2015;112(49):E6736–E6743.
9. Hammond A, Galizi R, Kyrou K, et al. A CRISPR-Cas9 gene drive system targeting female reproduction in the malaria mosquito vector *Anopheles gambiae*. *Nat Biotechnol*. 2016;34(1):78–83.
10. Hammond AM, Galizi R. Gene drives to fight malaria: current state and future directions. *Path Global Health*. 2017;111(8):412–423.
11. Moreira LA, Iturbe-Ormaetxe I, Jeffery JA, et al. A Wolbachia symbiont in *Aedes aegypti* limits infection with dengue, Chikungunya, and Plasmodium. *Cell*. 2009;139(7):1268–1278.
12. Ostera GR, Gostin LO. Biosafety concerns involving genetically modified mosquitoes to combat malaria and dengue in developing countries. *JAMA*. 2011;305(9):930–931.

13. Jinek M, Chylinski K, Fonfara I, Hauer M, Doudna JA, Charpentier E. A programmable dual-RNA-guided DNA endonuclease in adaptive bacterial immunity. *Science*. 2012;337(6096):816–821.

14. Anderson ME, Mavica J, Shackleford L, et al. CRISPR/Cas9 gene editing in the West Nile Virus vector, *Culex quinquefasciatus* say. *PLoS ONE*. 2019;14:1–10.

15. Chaverra-Rodriguez D, Macias VM, Hughes GL, et al. Targeted delivery of CRISPR-Cas9 ribonucleoprotein into arthropod ovaries for heritable germline gene editing. *Nat Commun*. 2018;9:1–11.

16. Dong S, Lin J, Held NL, Clem RJ, Passarelli AL, Franz AWE. Heritable CRISPR/Cas9-mediated genome editing in the yellow fever mosquito, *Aedes aegypti*. *PLoS ONE*. 2015;10(3):1–13.

17. Li M, Bui M, Yang T, Bowman CS, White BJ, Akbari OS. Germline Cas9 expression yields highly efficient genome engineering in a major worldwide disease vector, *Aedes aegypti*. *PNAS*. 2017;114(49):E10540–E10549.

18. Liu F, Ye Z, Baker A, Sun H, Zwiebel LJ. Gene editing reveals obligate and modulatory components of the CO_2 receptor complex in the malaria vector mosquito, *Anopheles coluzzii*. *Insect Biochem Mol Biol*. 2020;127(September):1–9.

19. Macias VM, McKeand S, Chaverra-Rodriguez D, et al. Cas9-mediated gene-editing in the malaria mosquito *Anopheles stephensi* by ReMOT control. *BioRxiv*. 2019;10(April):1353–1360.

20. Hall AB, Basu S, Jiang X, et al. A male determining factor in the mosquito *Aedes aegypti*. *Science*. 2016;348(6240):1268–1270.

21. Adelman ZN, Tu Z. Control of mosquito-borne infectious diseases: sex and gene drive. *Trends Parasitol*. 2016;32:219–229.

22. Fasulo B, Meccariello A, Morgan M, Borufka C, Papathanos PA, Windbichler N. A fly model establishes distinct mechanisms for synthetic CRISPR/Cas9 sex distorters. *PLoS Genetics*. 2020;16(3):1–22.

23. Sun H, Liu F, Ye Z, Baker A, Zwiebel LJ. Mutagenesis of the orco odorant receptor co-receptor impairs olfactory function in the malaria vector *Anopheles coluzzii*. *Insect Biochem Mol Biol*. 2020;127(September):332–345.

24. Zhu GH, Jiao Y, Chereddy SCRR, Noh MY, Palli SR. Knockout of juvenile hormone receptor, methoprene-tolerant, induces black larval phenotype in the yellow fever mosquito, *Aedes aegypti*. *PNAS*. 2019;116(43):21501–21507.

25. Oxitec. Our company and culture. Oxitec, LTD; 2021.

26. Gossen M, Bujard H. Tight control of gene expression in mammalian cells by tetracycline-responsive promoters. *Proc Natl Acad Sci U S A*. 1992;89(12):5547–5551.

27. Gong P, Epton MJ, Fu G, et al. A dominant lethal genetic system for autocidal control of the Mediterranean fruitfly. *Nat Biotechnol*. 2005;23(4):453–456.

28. Phuc HK, Andreasen MH, Burton RS, et al. Late-acting dominant lethal genetic systems and mosquito control. *BMC Biol*. 2007;5(1):11.

29. Fu G, Condon KC, Epton MJ, et al. Female-specific insect lethality engineered using alternative splicing. *Nat Biotechnol*. 2007;25(3):353–357.

30. Oxitec. Oxitec successfully completes first field deployment of 2nd generation Friendly *Aedes aegypti* technology. Oxitec, LTD; 2019.

31. Oxitec. Oxitec's Friendly™ mosquito technology recieves U.S. EPA approval for pilot projects in U.S. Oxitec, LTD; 2020.

32. Anonymous. First U.S. test for GM mosquitoes. *Science (News Focus)*. 2021;372:442.

33. Oxitec. State of Florida approves Oxitec experimental use permit. Oxitec, LTD; 2020.
34. Oxitec. Oxitec announces landmark approval of Florida Keys pilot project to combat mosquito that transmits dengue, zika. Oxitec, LTD; 2020.
35. Oxitec. Landmark project to control disease carrying mosquitoes kicks off in the Florida Keys. Oxitec, LTD; 2021.
36. Caragata EP, Dong S, Dong Y, Simões ML, Tikhe CV, Dimopoulos G. Prospects and pitfalls: next-generation tools to control mosquito-transmitted disease. *Ann Rev Microbiol.* 2020;74:455–475.
37. Godfray HCJ, North A, Burt A. How driving endonuclease genes can be used to combat pests and disease vectors. *BMC Biol.* 2017;15(1):1–12.
38. Carballar-Lejarazú R, James AA. Population modification of Anopheline species to control malaria transmission. *Path Global Health.* 2017;111(8):424–435.
39. Collins JP. Gene drives in our future: challenges of and opportunities for using a self-sustaining technology in pest and vector management. *BMC Proc.* 2018;12(Suppl 8):32–41.
40. Reegan AD, Ceasar SA, Paulraj MG, Ignacimuthu S, Al-Dhabi NA. Current status of genome editing in vector mosquitoes: a review. *BioSci Trends.* 2016;10(6):424–432.
41. Esvelt KM, Smidler AL, Catteruccia F, Church GM. Concerning RNA-guided gene drives for the alteration of wild populations. *eLife.* 2014;3(July2014):1–21.
42. Ledford H. CRISPR, the disruptor. *Nature (News Feature).* 2015;522:20–24.
43. Oye KA, Esvelt KM, Appleton E, et al. Regulating gene drives. *Science.* 2014;345(6197):626–628.

chapter three

Setting Up a Public Health Entomology Program

Structure, Organization, and Classification Issues

Every state-level public health department needs a medical or public health entomology group or section, consisting of one or more entomologists or public health biologists. This need may extend to city or county health departments as well, or pest abatement districts, especially along the coast or in areas with considerable vector-borne disease problems. The first thing that must be decided is where to put the entomologist(s), that is, what department, division, section, or branch. In the early days of public health departments, entomologists were mostly placed within sanitation departments or sanitary engineering groups,[1] but now, in most states, they are housed in the epidemiology/communicable disease group, the vector-borne disease group, the environmental health group, or some combination thereof. For example, in Tennessee, the public health entomologist position is located within Communicable and Environmental Disease Services; in Michigan, the zoonotic disease biologist is assigned to the Division of Communicable Disease; and in Mississippi, the medical entomologist is placed in the Office of Environmental Health.

The next thing that must be decided is what kind of person is right for the position. Ideally, this should be somebody with a wide variety of entomology experience, preferably public health entomology experience, but certainly medical entomology, possibly even veterinary entomology (herd health is the same no matter what species). At a very minimum, a zoologist or field biologist may suffice, as long as he or she has taken courses in medical entomology and understands the basics of insect identification and surveillance methods. Many state-level entomology programs require a master's degree or higher in entomology or medical entomology. A "job content" example of the knowledge, skills, and abilities required for a state-level public health entomology position is included in Table 3.1. Interestingly, one of the best groups from which to hire public health entomologists is the U.S. military. The U.S. Army, Navy, and Air Force have over 100 active-duty uniformed entomologists appropriately trained to recognize a wide variety of vectors and vector-borne diseases.[2] These entomologists are outstanding candidates for public health entomology programs.

Table 3.1 Sample Job Content Questionnaire for the Position of Medical or
Public Health Entomologist.

Job summary: Medical entomologist; prevention and management of arthropod-
caused or transmitted diseases.
Educational requirements: MS or higher in public health (medical) entomology;
some training in pesticide safety and usage.
Machines or equipment: Dissecting and compound microscopes; insect traps
and collecting equipment.
Working conditions: During fieldwork, some exposure to extreme, heat, cold,
other bad weather, and insect bites/stings.
Special requirements: A person should be reasonably mobile and physically fit
to do on-site investigations.
Duty statement 1: Provide technical advice and consultation on all aspects of
medical biology/zoology/entomology to all local, district, and state-level health
departments, and state or local public health or preventive medicine personnel.

Amount of time devoted to this duty: 40%	How frequently is this duty performed? Regularly	Consequence of error (1–5): 3

Tasks: Answering written or telephone requests for information about insects,
insect-borne diseases, or other pest problems. Looking up information (upon
request) online, in reference books, personal scientific papers, or medical school
libraries.
Knowledge required: Knowledge of all aspects of medically important insect
pests and their management and control.
Skills required: Speaking and writing skills to convey up-to-date
recommendations for management of pest problems.
Abilities: Ability to handle sensitive (controversial at times) pest problems.
Ability to translate technical information into layperson's terms.
Duty statement 2: Collect and identify arthropods throughout the state with
special emphasis on those that may pose health problems (so as to determine
which harmful insects occur in the state).

Amount of time devoted to this duty: 30%	How frequently is this duty performed? Regularly	Consequence of error (1–5): 3

Tasks: Collect specimens by various means in a variety of locations. Identify
specimens using published scientific articles and books. Maintain a reference
collection of insect specimens.
Knowledge required: Knowledge of insect life cycles, morphology, and
taxonomy.
Skills required: Use of an identification key, site selection for field sampling
equipment, and some manner of filing/storing reference insects.
Abilities required: Ability to sit for long periods of time working through
taxonomic keys (for identification) and looking through a microscope.

Duty statement 3: Review research data, read, and study scientific literature, in order to train or provide consultation to health department personnel in insect control methodologies and pesticide/environmental health issues.

Amount of time devoted to this duty: 10%	How frequently is this duty performed? Infrequently	Consequences of error (1–5): 3

Tasks: Keep abreast of changes in related science and technology. Conduct seminars, in-services, and workshops on current pest control practices, pesticides, and future trends.

Knowledge required: Knowledge of mosquito, tick, and rodent biology, ecology, and control. Knowledge of complex scientific literature.

Skills required: Speaking and writing skills to relate complex information to laypersons.

Abilities required: Ability to convey the need for changes in a nonthreatening manner.

Duty statement 4: Provide information on mosquito control techniques to local municipal mosquito control agencies or mosquito abatement districts.

Amount of time devoted to this duty: 10%	How frequently is this duty performed? Infrequently	Consequence of error (1–5): 3

Tasks: Telephone, e-mail, or in-person visits to municipal mosquito control shops or mosquito abatement districts to discuss their program(s). Conduct workshops on safe and effective methods for mosquito control.

Knowledge required: Knowledge of pesticides, equipment, and techniques used in mosquito control and EPA and state laws governing pesticides and the environment.

Skills required: Strong oral and written communicative skills.

Abilities required: Ability to relate to people of varying ethnic, political, or socioeconomic backgrounds.

Duty statement 5: Design and frequently revise printed and online material about setting up municipal mosquito control programs, and ensure that state, local, and municipal personnel have this information at their disposal.

Amount of time devoted to this duty: 5%	How frequently is this duty performed? Infrequently	Consequence of error (1–5): 3

Tasks: Devise, print up, frequently revise, and distribute guidance about suppressing mosquito populations in areas of the state where mosquito-borne disease outbreaks occur.

Knowledge required: Knowledge of mosquito-borne diseases in the state. Knowledge of mosquito control methodology, equipment, and chemicals.

Skills required: Writing skills to write and revise large documents or online postings.

Abilities required: Ability to communicate in written form the importance of mosquitoes and mosquito control.

(Continued)

Table 3.1 Sample Job Content Questionnaire for the Position of Medical or
Public Health Entomologist. (Continued)

Duty statement 6: Performs other duties when requested or assigned.		
Amount of time devoted to this duty: 5%	**How frequently is this duty performed: Infrequently**	**Consequence of error (1–5): 3**

Tasks: Performs other duties as assigned.
Knowledge required: Knowledge of agency policies and procedures. Knowledge of applicable regulations and laws.
Skills required: Strong oral and communication skills.
Abilities required: Ability to get along with others. Ability to follow directions.

Once an entomologist is selected, the state or local health officer, along with the state epidemiologist or public health veterinarian, need to define and clarify for the entomologist the extent to which he or she is to work with other groups within the health department. For example, public health entomologists interact with pest control and pesticide regulatory groups like the department of agriculture (in Mississippi this is the Bureau of Plant Industry); however, entomologists are also involved with disease investigations along with epidemiology or infectious disease personnel or even with public health veterinarians. The public health entomologist should be a "people person," able to work with a wide variety of people in all sorts of different situations. Often, during an investigation, the entomologist is accompanied by an epidemiologist or nurse to answer clinical questions, while the entomologist is there to answer environmental or entomological questions. The entomological and environmental aspects should not overlap the clinical aspects of a case. Clinical personnel are appropriately trained, licensed, and insured to discuss those aspects of human illness; entomologists are not.

In many cases, the public health entomologist is assigned to the division of environmental health, previously called the sanitation department, although this does not mean garbage collection. Sanitarians, environmentalists, or environmental health specialists are the people in health departments who inspect restaurants, day care centers, and health care facilities, as well as those responding to general environmental sanitation complaints and other environmentally related health issues (Figure 3.1). Public health entomologists should appreciate the fact that these environmental health specialists perform a variety of functions and are foot soldiers in public health. They can be the entomologist's hands, eyes, and feet in the field (Figure 3.2).

Figure 3.1 Environmental health specialists perform inspections at food establishments, day care centers, and dairy facilities.

Source: Photo courtesy Dr. Jeffrey Brown, Mississippi Department of Health

Dipping for larvae   Transferring larvae into vials

Taking environmental samples   Dragging for ticks

Figure 3.2 Environmental health specialists may assist in a variety of vector control functions.

Source: Photo courtesy the Centers for Disease Control

Clarification of the Public Health Entomologist's Role

Private versus Public Pest Problems

Governmental or public pest or vector control is obviously different from private pest problems, and public health entomologists need to be aware of this and be able to rightly distinguish between the two (Table 3.2). In the United States, a person's private pest issues in or around his or her home, such as fleas, ants, or cockroaches, are strictly his or her responsibility, as well as paying for their extermination. The government or health department will not come do this or pay for it to be done. Of course, there can be overlap between private and public pest problems, wherein a private fly, ant, or rodent problem becomes a nuisance affecting others in the entire community. In that case, public health officials may take enforcement action to abate the nuisance and protect the public.

Public health pest control may be viewed differently in other countries. For example, in the United Kingdom, provisions of the Public Health Act (1936, revised 1961) empower local authorities and school authorities to deal with "verminous" conditions of either people, houses, or movable articles. This may be done at the expense of the local authority. Pest hygiene issues are generally handled by an environmental health officer (EHO) and may involve cleaning and destroying verminous articles or

Table 3.2 Some Examples of Private versus Public Health Problems.

Private pest problems	Public pest problems	Both (overlapping public and private)
Insect pests such as ants, cockroaches, fleas, bed bugs, in a private home or office	Insect or rodent pests such as ants, cockroaches, fleas, bed bugs, or rats in public places such as schools, hotels, hospitals, nursing homes, jails, residential shelters, and public transit	Insect or rodent pests such as ants, cockroaches, fleas, bed bugs, or rats in public housing units, a few houses throughout a neighborhood or in an apartment complex
Commensal rodents (rats and mice) in a private home or office	Widespread pest infestations throughout a community (e.g., entire neighborhoods)	
Bats in a private home or office	Bird roosts	

Note: Other countries in the world may interpret this differently, and in those cases, public health authorities may be involved in abatement of private pest issues.

cleansing of verminous properties. The local authority may serve notice to the owner of a property to have the place properly exterminated and, if necessary, may have it done, and recover the cost. As a last resort, legal action may be taken against property owners, tenants, or food establishments for various pest infestations. This may be done by local authorities through their EHOs and entomologists may be called as expert witnesses.

In the United States, the public health entomologist is contacted about various pest problems and nuisances. Knowing when to go perform an inspection or environmental assessment is critical for wise time management. Many things can be handled over the phone. For example, if someone calls about ants in their kitchen or bats in their attic, the PHE can politely answer their questions, even perhaps offering general comments about their control, but ultimately, the complainant will have to get a private pest control company to deal with the issue(s). However, if someone calls with complaints about pests in a school, or nursing home, or over a wide area like entire neighborhoods, that constitutes a public health nuisance that requires inspection and possible intervention.

Duties of the Public Health Entomologist

In the early to mid-20th century, health department entomologists performed insect identification, control, and disease prevention duties, mostly related to malaria or filth fly—associated diseases.[1] Not much has changed, except for the specific diseases. Twenty-first century public health entomologists often identify entomological specimens sent in from the public or the medical community and give advice as to control or management of these pests (see Table 3.1 and Figure 3.3). Analysis of my logbook when at the Mississippi Department of Health from 2002 through 2006 revealed several key issues in public health entomology (Figure 3.3). Eighty percent of all calls and e-mails were about general vector complaints and questions, including bird and bat roosts; rodent problems; and mosquito, fly, or tick problems around the state. Many such queries were about ways to control these pests. Specimen identification was the second-largest category of inquiries (total 32%), with 22% of consults being specimens sent in from the public and 10% specimens submitted by medical professionals. Answering questions and consultations about disease outbreaks such as West Nile virus (22%) constituted another large portion of time as a public health entomologist. Interestingly, a disproportionate amount of time (16%) was spent dealing with people who thought that insects or mites were living on them, a condition called Ekbom syndrome or delusions of parasitosis (DOP).[3] This is consistent with published accounts of DOP sufferers making multiple contacts with physicians, parasitologists, public health departments, and other governmental agencies.

SECTION FROM AN ACTUAL EMPLOYMENT ADVERTISEMENT FOR A PUBLIC HEALTH ENTOMOLOGIST

As Public Health Entomologist, this position is responsible for developing and maintaining surveillance of vectors of arthropod-borne diseases. This position is also responsible for providing expert advice and consultation to local health departments and state agencies on identifying and controlling the vectors of diseases in the state, including appropriate use of pesticides and repellents.

Performing arthropod or vector-borne disease surveillance is another important component of public health duties (see Chapter 4) and involves both collecting arthropod specimens and coordinating the collection of blood or other useful tissues/samples. Another duty in the public health entomology job description is to train the environmental health specialists in pest and pest control issues (see the next two sections). Well-trained environmental health specialists or inspectors are critical to the overall mission of public health, and the entomologist is responsible for the insect-related aspects of his or her training. Also, these environmental health specialists are often the ones who bring in the samples to be identified and processed by the public health entomologist. The entomologist may also perform mosquito and other vector surveillance and provide those data and educational training to mosquito abatement district (MAD) personnel located throughout his or her state. Some states, like Florida or Louisiana, have many MADs, while others have none. Another thing the entomologist does is to give media interviews and write articles for scientific journals or popular press outlets, such as magazines, and also give speeches about entomology and bug-related health issues at garden clubs, schools, or other civic organizations. These extension-like efforts are aimed at educating the public about the importance of insects and other arthropods in human health. Also, as mentioned earlier, entomologists need to be available for vector-borne disease outbreak investigations, especially to accompany clinical staff while looking at cases of human disease outbreaks caused by arthropods.

TIPS FOR AVOIDING LEGAL ISSUES

(See Also Discussion in Chapter 5)
Over the past few decades there has been a tremendous increase in litigation regarding arthropods—especially cases resulting from

specimens found in hotels (such as bed bugs), food products, and so on. There are even people suing cities for fire ant stings received while visiting the city park. Supposedly, the park should have had an "appropriate" fire ant control program. Due to this increase in litigation, proper documentation and record keeping are necessary since many insect samples submitted for identification become integral parts of lawsuits. Persons finding a "bug" or "maggot" in their food are often advised by their attorney to have the event documented and the specimen identified, hence the basis for submitting the sample to the public health entomologist (PHE). PHEs should assume a lawsuit is forthcoming and act accordingly. The following tips may be helpful:

1. Make a careful identification or submit the specimen to a university specialist for that particular group of arthropod.
2. Document all aspects of the sample (who, what, when, where). It's helpful to include lot numbers on food products.
3. Use an ink pen and a bound logbook with consecutively numbered pages (it's more difficult to argue that someone tampered with the data).
4. Retain the specimens in alcohol or in a freezer for at least 1 year.
5. Do not stray beyond your area of expertise. Do not comment on the life cycle, ecology, or health effects of a particular insect unless you are reasonably sure of the facts. Such speculation can come back to haunt you.

The entomologist needs to be an expert in pesticides. As discussed in the previous chapter, pesticides are important public health tools to manage out-of-control populations of insects and disease vectors. We live in a chemophobic society, so people are generally not amenable to indiscriminate pesticide use; however, clinical staff and epidemiologists understand that pesticides may need to be used sometimes to stop vectorborne diseases. The entomologist is the primary resource person in this regard, understanding what pesticides are available currently and properly registered by the EPA. When I was at the Mississippi Department of Health, about 10% of my calls and e-mails were about pesticides and their proper uses (Figure 3.3). Although the public health entomologist may not actually apply pesticides, he or she should have commercial applicator certification in category 8, public health pest control. A *license* in this category is not necessary, but *certification* is highly recommended. Federal law

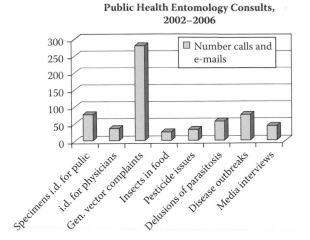

Public Health Entomology Consults, 2002–2006

Figure 3.3 Analysis of logbook entries made by a public health entomologist from 2002 through 2006 at the Mississippi Department of Health.

requires that anyone using "restricted use" pesticides must be certified or under the direct supervision of someone who is. Direct supervision means that the certified applicator must be readily available to give directions or advice, although usually he or she does not have to be present at the site of treatment.

Another reason the entomologist should be certified is that he or she may sometimes help in procurement of pesticides and advise how to use them in a safe and effective manner (see text box about potential conflicts of interest). The entomologist should be able to coordinate with local, county, city, and municipal mosquito control personnel, or other vector control groups such as mosquito control districts, to appropriately select and use these products in mosquito or other vector control. For example, in some outbreaks of mosquito-borne encephalitis, the only way to block or prevent further spread of the disease is to kill the mosquito vectors, and appropriate pesticides need to be chosen in consideration of habitat, environmental conditions, extent of human exposure, and other such factors.

The public health entomologist may be involved to some extent in basic research on insects and other arthropods of medical importance (see Chapter 7). Obviously, these efforts should be coordinated with the health officer or epidemiologist to ensure their relevance to the mission of public health (for example, the entomologist shouldn't be conducting research on the best ways to kill sweet potato weevils). Basic research, and especially survey type studies to determine what species of ticks, fleas, mosquitoes, and so on, occur in the state, where they occur, and when they are active, can be a contributing factor to the overall success of the program

Figure 3.4 Public health entomologist, Afsoon Sabet, identifying mosquitoes.

(Figure 3.4). If the entomologist never goes out in the field to determine what arthropod pests occur in his or her area, and when they are active, then he or she will be ill-equipped to answer calls and questions about these pests.

POTENTIAL CONFLICTS OF INTEREST

State or federal employees are public servants and should be careful to avoid conflicts of interest. Most states specifically prohibit use of one's office or position for monetary gain other than that allowed by law. For example, Mississippi Code § 25–4–1 states, "The Legislature hereby declares it essential to the proper operation of democratic government that public officials and employees be independent and impartial; that governmental decisions and policy be made in the proper channels of the governmental structure; that public office not be used for private gain other than the remuneration provided by law; that there be public confidence in the integrity of government; and that public officials be assisted in determinations of conflicts of interest." Further, state employee handbooks often make statements such as, "Employees and public servants should be especially careful to avoid using, or appearing to use, an official position for personal gain, giving unjustified preferences, or losing sight of the need for efficient and impartial decision making." In light of these ethics rules, public health entomologists should be very cautious in recommending a particular brand of pesticide or commercial product such as traps or devices for pest control. Instead, he/she should always try to recommend only active ingredients and/or provide a list of several branded products.

Lastly, the public health entomologist should be keenly aware of what his or her role is *not*. This includes examining patients and offering clinical advice. In rural areas, citizens may look to any member of the public health department as a "healthcare provider" and ask them clinical questions. Entomologists should not fall into this trap and need to be careful not to stray beyond their entomological experience. One particular example is the case of delusions of parasitosis wherein people are convinced that tiny mites or insects are on or in their skin.[4,5] These patients frequently have a long history of doctor visits and pest exterminator visits and are so desperate for help that they will ask entomologists to examine their lesions (anywhere on the body) or discuss the clinical aspects of their condition. This is unethical on several fronts, and invariably will backfire on the entomologist when the patient fails to get relief from the invisible bugs.

Providing Entomology Training

Health department entomologists may provide training to physicians, nurses, family nurse practitioners, physician assistants, and other healthcare providers on all aspects of vectors and vector-borne diseases. One avenue for such training is continuing medical education events coordinated through a local medical school or hospital. For example, I was involved in developing and conducting 1-day continuing medical education (CME) courses titled "Entomology for the Medical Practitioner," held twice (January 18, 1995 and October 29, 1999) at the University of Mississippi Medical Center in Jackson. Announcements and registration forms for the event were sent statewide to physicians in three specialties (family practice, dermatology, and emergency medicine), as well as to nurse practitioners, medical technologists, and toxicologists (poison control center personnel). The first event in 1995 consisted of three lectures on medical entomology, a case studies session, and a laboratory session. The 1999 event was modified to include an additional entomology lecture and a treatment lecture taught by a physician from a local emergency department. The topics of the medical entomology lectures covered (1) direct effects, such as bites and stings; (2) indirect effects, such as disease transmission and insect allergies; (3) lesions caused by arthropods; and (4) ticks and tick-borne diseases. The case studies session was essentially a group activity, involving scenarios of real or imagined arthropod problems. The group was challenged to determine possible causes and remedies. Laboratory sessions included demonstration of microscopic and gross specimens of all of the major arthropod parasites of humans. "Unknown" specimens were provided for the students to identify, and self-tests were administered. In both cases, the textbook used was *Physician's Guide to Arthropods of Medical Importance*[6] (now called *Goddard Guide to Arthropods of Medical Importance*).[7] Other medical entomology texts, such as *Medical Entomology for Students*,[8] could easily be substituted, however.

VARIOUS GROUPS THE PUBLIC HEALTH ENTOMOLOGIST MIGHT HELP TRAIN

- Health department inspectors (environmental health specialists)
- Physicians (continuing medical education events)
- Laboratory technicians (continuing medical education events)
- Municipal mosquito and vector control personnel
- Mosquito abatement district personnel

Both CME events were well attended and received excellent evaluations regarding their form, structure, and relevance. The sessions in 1995 and 1999 drew 16 and 12 participants, respectively, and results of the 1999 course evaluations revealed that all participants rated the events as "excellent" meetings that were "about right" in length (Table 3.3). The majority of respondents said that the activity enhanced their current medical knowledge.

Table 3.3 Evaluation Results for Healthcare Provider CME Event (1999)
Presented by a Public Health Entomologist.

Question asked	Response	Percent marking that response
What overall rating would you give the entire meeting?	Excellent	100%
Was the length of the meeting:	About right	100%
To what degree did this activity enhance your current knowledge?	Very much	58%
	Moderate	42%
To what degree will you use the information from this activity in your clinical practice?	Very much	42%
	Moderate	50%
Were your personal objectives for attending this course satisfied?	Very much	92%
	Moderate	8%

While many of the most serious insect-related health problems and vector-borne diseases do not ordinarily occur in the United States (although some do), international travel has created a global village in which tropical maladies are easily imported. Most recently, increases in "adventure vacations" and ecotourism have contributed to the spread of tropical infections. Accordingly, there is an ongoing need for physician training in medical entomology. Perhaps CME events such as the ones described here can serve as a model for use by other state health departments.

Medical Student and Laboratory Technician Training

Sometimes the public health entomologist helps train medical students or laboratory personnel about arthropods and health. There is certainly an increasing need for this training. Medical student training (the 4-year standard course) has changed dramatically since World War II. The amount of basic biological information that each medical student is required to master in the preclinical curriculum has vastly increased; coverage of many traditional, classic disciplines has therefore been dropped to make way for voluminous information in rapidly expanding areas such as molecular biology, genetic testing/counseling, neuroscience, and pharmacology.[9,10] Examples of classic disciplines dropped to make way for new ones include parasitology and medical entomology. The Association of American Medical Colleges maintains a detailed comparison of curricula of participating American and Canadian medical schools as an online searchable named CurrMIT.[11] A search of CurrMIT using the keyword *arthropods* revealed that only 22 of 126 participating medical schools included arthropods in any of their course content. For most of these institutions, information on arthropods comprises only a small part of a course in microbiology or pathology.

My personal experience with medical student training resulted from a clinical assistant professor appointment (now called affiliate faculty) at the University of Mississippi Medical School, which allowed me to teach the arthropods section of the course, "Medical Parasitology," to second-year medical students (M2s). Originally, this consisted of two lectures and a lab concerning arthropods of medical importance that medical practitioners might encounter in practice. However, after a few years, the "Medical Parasitology" course was eliminated, its contents truncated, and portions of it placed within a larger course called "Medical Microbiology." Under this new combination, I continued to teach M2s about medically important arthropods, though now only in two lectures. This two-lecture format continues to this day. While this (2 h) seems woefully inadequate for entomological training of future doctors, it is nonetheless better than what is available at many medical schools, where little or no medical entomology is contained in the curriculum.

Perhaps designers of medical school curricula may be assuming that information pertaining to the identification of disease-causing arthropods is a component of the training of medical laboratory technologists/clinical laboratory scientists. Two organizations that administer national examinations to certify clinical laboratory scientists, however, do not usually include medical entomology as part of the examination content. The American Society of Clinical Pathologists (ASCP) has recommended that material related to arthropods should be "deleted or should not be included in the curriculum" for medical technology.[12] ASCP provides detailed examination content guidelines for all specialties in the clinical laboratory sciences. One search of these guidelines revealed that the term *arthropods* was not even mentioned for the categories medical laboratory technician (MLT) and medical laboratory scientist (MLS).[13,14] For the categories technologist in microbiology (M) and specialist in microbiology (SM), the guidelines say that 15% of the exam concerns parasitology, with about 1/8 of that being arthropods.[15] Obviously, medical entomology is not considered very important in the knowledge base of an entry-level clinical laboratory scientist. The question then becomes: Which, if any, healthcare professional is being trained in medical entomology? Who is going to identify arthropod specimens for the medical community?

Unless another major war is fought in tropical regions, it is unlikely that courses in medical entomology and parasitology will return to medical training. Realistically, the best that can be hoped for is inclusion of medical entomology within medical microbiology courses, environmental medicine courses, or infectious diseases training. There is, however, a definite need for broader coverage of these subjects. The importance of entomology and parasitology in healthcare worker training is highlighted by recent examples of vector-borne tropical diseases being introduced into the urban environment by rapid modern air travel. The latest major example of this was introduction of West Nile encephalitis into the United States.[16]

Despite this lack of entomology training for medical professionals, a large portion of tropical diseases is caused by or transmitted by arthropods and parasites,[17] and approximately 60% of all known human pathogens are zoonotic.[18] Immigration, both legal and illegal, combined with rapid international air travel, increasingly bring these diseases into industrialized countries.[19] The number of international departures from airports in the United States doubled in a 12-year span, climbing from 20 million in 1983 to nearly 40 million in 1995; more than 50% of those departures each year were bound for tropical countries.[20] This level of international travel greatly increases the risk of introduction of foreign disease agents. In addition, adventure travel or "eco-vacations" are the largest growing segment of the leisure travel industry, with growth of 10% per year since 1985,[21] so people are bringing exotic pests and diseases back from these tropical locations. Who is charged with recognizing these arthropod or worm pests in the clinical setting? It is apparently not physicians or laboratory technologists. The author often gets entomologic specimens from laboratory personnel (specimens originally sent from physicians) who do not even know to which major group the organism belongs. For example, once a soft tick (family *Argasidae*) was removed from a patient's ear by a physician who called it a "baby crab." Additional lab technician and physician training is desperately needed. Just as is the case with microbial agents, treatment or control recommendations for arthropod problems hinge on correct identification, a skill that is best acquired with hands-on training.

Sanitarian or Environmental Health Specialist Training

A survey of state health department vector control directors found that 74% did not have sufficient numbers of public health workers to effectively staff their vector control units.[18] Therefore, environmental health specialists (EHSs) must be trained to help fill this gap. Virtually every state, county, or local health department has inspectors or EHS personnel who conduct food, dairy, wastewater, day care, and other general environmental health regulatory duties. Privy construction, dairy inspections, and malaria control were among the first activities of these workers (Figure 3.5). In 1929, Dr. Felix Underwood, state health officer of Mississippi, initiated efforts to make malaria control an integral function of county health departments.[22] This was among the first efforts nationwide to establish trained sanitarians in every county to investigate diseases, perform inspections, and conduct general sanitation education and investigations.

Figure 3.5 Historically, county health departments were involved in basic sanitation activities such as construction of privies.

Source: From the National Library of Medicine

The public health entomologist is charged with training these county inspectors or EHS staff about entomology and pesticide-related subjects. This EHS training is often the only entomological education that inspectors receive, so it must be done carefully and thoroughly. Since EHSs are foot soldiers of public health, being the first line of defense against diseases arising from the environment, they require broad training. The public health entomologist may choose to set up scheduled, periodic training for EHSs or conduct training on an as-needed basis. At a minimum, entomological training should be included in "new environmental health employee" training and orientation conducted upon first employment with the agency. In Mississippi, we tried to combine entomology education (1–3 h) with other training sessions such as food, water, or milk safety. That strategy minimized EHS time away from their regular duties. All such efforts should be coordinated with the agency training/technical director who schedules and documents employee training.

Other training is available at the national level. The National Environmental Health Association (NEHA) has partnered with the CDC National Center for Environmental Health to offer workshops titled "Biology and Control of Vectors and Public Health Pests" at NEHA annual conferences and also at several regional sites.[23] These workshops are aimed at training EHSs in the following core areas:

1. The identification, biology, and control of insects and rodents of public health importance
2. Commonly used insecticides in the control of insects of public health importance

3. Commonly used rodenticides in the control of rodents of public health importance
4. Integrated pest management (IPM) techniques in the control of insects and rodents of public health importance
5. Understanding zoonotic diseases of public health importance that are transmissible to humans from insects and rodents and the associated risks of transmission to humans

In addition, NEHA has made this and many other training materials available online with opportunities for CME credits. For more information, search for courses under professional development and training and vectors and pest control at www.neha.org.

Role of Public Health Veterinarians in Entomological Training

Many state health departments have State Public Health Veterinarians (SPHV). These individuals are different from a "state veterinarian" (SV), who works for the state agriculture department.[24] SPHVs generally work in zoonotic disease control and prevention—diseases transmitted from animals to people, while SVs primarily target livestock diseases (some of these may be zoonotic) and their activities are primarily concentrated on benefits and protection to livestock and the livestock industry. SPHVs focus on protecting public health, and thus, as might be expected, over 70 public health veterinarians are employed at the CDC.[24] Credentialing and Board Certification for public health veterinarians is mainly governed through the American College of Veterinary Preventive Medicine, but some SPHVs are board-certified as parasitologists, pathologists, in laboratory animal medicine and other disciplines within the profession. There is also an American Association of Public Health Veterinarians, American Association of Wildlife Veterinarians, and the Alliance of Veterinarians for the Environment that address public health concerns.

Public health veterinarians in state health departments are usually housed in Epidemiology or Communicable Diseases Departments, but may be assigned to Environmental Health. SPHVs are an integral part of training at health departments, both for clinical (doctors and nurses) and environmental staff (sanitarians and environmental health specialists). Their specific expertise for training programs generally falls into the areas of antimicrobial resistance, animal bites, rabies, other zoonotic diseases, parasites, vectors, and the (animal) hosts of zoonotic diseases. For example, the SPHV might be needed to train entomology and environmental personnel as to proper wild bird trapping and bleeding techniques for arboviral surveillance. In addition, they may provide guidelines for animals in schools, health care facilities, and as service and therapy animals.

In addition, SPHVs are the local and state professionals who regularly consult with physicians, emergency rooms, legislators, local officials, schools, health departments, and the general public on preventing exposures to and controlling diseases that humans can get from animals and their products.[24] Many SPHVs are on call continuously, especially in relation to rabies exposures. No list of local or state officials can be considered complete without the SPHV and local public health veterinarians.

References

1. Eads RB, Irons JV. Current status of public health at the state level. *Am J Public Health Nations Health*. 1951;41:1082–1086.
2. Cope SE, Schoeler GB, Beavers GM. Medical entomology in the United States Department of Defense: challenging and rewarding. *Outlooks Pest Manag*. 2011;June:129–133.
3. Hinkle N. Ekbom syndrome: the challenge of "invisible bug" infestations. *Ann Rev Entomol*. 2010;55:77–94.
4. Goddard J. Imaginary insect or mite infestations. *Infect Med*. 1998;15:168–170.
5. Hinkle N. Delusory parasitosis. *Am Entomol*. 2000;46:17–25.
6. Goddard J. *Physician's Guide to Arthropods of Medical Importance*. 6th ed. Boca Raton, FL: Taylor and Francis (CRC); 2013.
7. Moraru GM, Goddard JI. *The Goddard Guide to Arthropods of Medical Importance*. Boca Raton, FL: CRC Press; 2019.
8. Service MW. *Medical Entomology for Students*. 5th ed. Cambridge, UK: Cambridge University Press; 2012.
9. Downie R, Charlton B, Calman K, McCormick J. *The Making of a Doctor*. New York: Oxford University Press; 1992.
10. Pope A, Rall D. *Environmental Medicine: Integrating a Missing Element into Medical Education*. Washington, DC: National Academy Press; 1995.
11. AAMS. Medical school curriculum management and evaluation website (currMIT); 2011. www.aamc.org/meded/curric/start.htm (accessed August 10, 2011).
12. ASCP. Technical curricula for MT and MLT training programs; 2001. www.ascp.org/bor/directors/tech_mt (accessed September 6, 2001).
13. ASCP. Medical laboratory technician (MLT) examination content guideline; 2009. www.ascp.org/pdf/BOR-PDFs/Guidelines/ExaminationContentGuidelineMLT.aspx.
14. ASCP. Medical laboratory scientist (MLS), examination content guidelines; 2009. www.ascp.org/pdf/BOR-PDFs/Guidelines/ExaminationContentGuidelineMT.aspx.
15. ASCP. Technologist in microbiology (M) and specialist in microbiology (SM), examination content guidelines; 2009. www.ascp.org/pdf/BOR-PDFs/Guidelines/ExaminationContentGuidelineSM.aspx.
16. Peterson LR, Roehrig JT. West Nile virus: a reemerging global pathogen. *Emerg Infect Dis*. 2001;7:611–614.
17. Goddard J. Arthropods and medicine. *J Agromed*. 1998;5:55–83.
18. Herring ME. Where have all the vector control programs gone? *J Environ Health*. 2010;73:30–31.

19. Cunningham NM. Lymphatic filariasis in immigrants from developing countries. *Am Fam Phys.* 1997;55:119–1204.
20. Gubler DJ. Arboviruses as imported disease agents: the need for increased awareness. *Arch Virol.* 1996;11:21–32.
21. Chomel BB, Belotto A, Meslin FX. Wildlife, exotic pets, and emerging zoonoses. *Emerg Infect Dis.* 2007;13:6–11.
22. Underwood FJ. The control of malaria as a part of full-time county health department program. *South Med J.* 1929;22:357–359.
23. Herring ME. Where have all the vector control programs gone, part two. *J Environ Health.* 2010;73:24–25.
24. Johnston WB. About public health veterinarians. In: NASPHV, "About Us"; 2010. www.nasphv.org/aboutPHVs.html.

chapter four

Vector-Borne Disease Surveillance

Overview of Surveillance Types

Until the 1950s, the word *surveillance* was only applied to a *person* with a disease entity such as plague and involved appropriate community alertness to keep that person quarantined and away from other, noninfected people. After World War II, that definition was changed with the formation of the U.S. Communicable Disease Center (now the Centers for Disease Control) to mean a *disease* or *condition* instead of a sick person.[1] This change fundamentally altered the way public health departments conducted disease surveillance. Under this new paradigm, public health entomologists are often charged with surveillance for vector-borne diseases (VBDs), which may include collecting arthropod specimens and testing them for various disease agents such as Lyme disease spirochetes, spotted fever group rickettsiae, ehrlichial organisms, and trypanosomes, among others. However, in most state public health programs, surveillance is primarily focused on mosquito-borne arboviruses such as those causing encephalitis. There is no one-size-fits-all arbovirus surveillance model. Some states have no vector surveillance of any kind, with little hopes of establishing one.[2] Therefore, in each state, surveillance systems must be tailored according to the probability of arbovirus activity and the reality of available resources. In states without any preexisting VBD programs, developing a new avian-based or mosquito-based arbovirus surveillance system may be required. For some projects, previously abandoned surveillance systems may need to be resurrected. In others, modification or strengthening of existing systems (e.g., for detection of West Nile virus (WNV), eastern equine encephalitis (EEE), western equine encephalitis (WEE), or St. Louis encephalitis (SLE)) may be the most appropriate response. States with low probability of arbovirus activity or lack of resources to support avian-based or mosquito-based surveillance may just conduct hospital or clinic surveillance for neurologic disease in humans and equines. Free-access information on how to establish surveillance programs is available online. Examples include the *CDC Guidelines for Arbovirus Surveillance Programs in the United States,*[3] *CDC Epidemic/Epizootic WNV in the United States,*[4] and the WHO chapter titled "Entomology in Public Health Practice."[5] The following discussion utilizes information contained in those documents.

Appropriate response(s) to surveillance data is the key to preventing human and animal disease associated with mosquito-borne arboviruses. That response must be immediate and effective mosquito control in the local area if virus activity is detected in the bird, mosquito, or human surveillance systems.

HEALTH DEPARTMENT VECTOR RELATED INFRASTRUCTURE

To protect against the risk of serious vector-borne disease (VBD) outbreaks, every state health department needs a functional VBD surveillance and response unit, staffed by well-trained personnel who have adequate data processing resources, suitable laboratory facilities, and an adequate operating budget. The size and complexity of these units may vary by state or territory, depending on (1) the importance and prevalence of VBD in the area and (2) available resources. A functional VBD surveillance unit should be considered an essential component of any emerging infectious diseases program. Ideally, VBD surveillance involves epidemiologists, virologists, medical entomologists, wildlife biologists, veterinarians, and data managers such as statisticians. The exact combination of personnel needed to conduct VBD surveillance depends on the importance of VBD in the area and available resources. In many health departments, a chronic shortage of medical entomologists exists. If possible, addressing this deficiency should be a high priority.

Sentinel Birds

Sentinel bird flocks can be used to detect arboviral disease in a local area. The ideal avian sentinel for arboviruses such as WNV and other encephalitis viruses would meet the following criteria: (1) universal susceptibility to infection, (2) 100% survival from infection, as well as universal development of easily detectable antibodies, (3) no risk of infection to handlers, and (4) no development of viremia sufficient to infect local vector mosquitoes. There is no perfect sentinel species, but domestic chickens and ducks have been used extensively and effectively as sentinels in many surveillance programs for WNV, SLE, EEE, and WEE viruses throughout the world (Figure 4.1). The suitability of chickens as sentinels has been demonstrated by showing that they can become infected in nature and also experimentally, after inoculation with small doses of virus, can become infected and circulate hemagglutination inhibition antibodies at detectable levels for at least a year.[6] Sentinel birds have never been shown

Figure 4.1 Bleeding a duck for Japanese encephalitis surveillance testing.

Source: Armed Forces Pest Management Board photo by David Hill

to pose an increased risk of arbovirus infection to their handlers or the human population at large. Monitoring farm and yard chickens for antibodies can be used in addition to, or as an alternative to, sentinel chicken-based arbovirus surveillance programs.

Advantages of Using Chickens and Other Birds

1. There is a long history (>8 decades) of successful use of chickens in flavivirus and alphavirus surveillance.
2. During the 1999 WNV outbreak, chickens, geese, and pigeons sampled in Queens Borough, New York City, all had a high seroprevalence (e.g., >50% prevalence of neutralizing antibody to WNV virus), demonstrating their usefulness as sentinels.
3. Chickens and other birds are readily fed upon by *Cx. pipiens, Cx. quinquefasciatus,* and *Cx. tarsalis* mosquitoes, which are primary vectors of WNV and SLE viruses.
4. Since chickens and other birds are kept in captivity, the geographic location of infection is never in question.
5. Chickens are relatively easy to bleed.
6. Collection and handling of specimens (serum) is relatively inexpensive.
7. No necropsies are needed.
8. The system is flexible and can be expanded and contracted as appropriate.

9. Mosquito abatement districts already have personnel and infra-structure to maintain flocks, bleed birds, and submit specimens to public health labs for testing.
10. Laboratory expenses can be defrayed by charging nominal fees per test.

Disadvantages of Using Chickens and Other Captive Birds

1. When sentinel flocks seroconvert, especially in regard to WNV, human cases are *already occurring* in the area. There is no lead time; mosquito larviciding, spraying, and public health educational campaigns must start immediately.
2. Sentinel flocks detect only focal transmission, requiring that multiple flocks be positioned in representative geographic areas.
3. Flocks may be subject to vandalism and theft, limiting their usefulness in urban areas.
4. Setup and flock maintenance can be expensive (i.e., birds, cages, feed, transportation).

Wild Bird Surveillance

Since WNV, EEE, and SLE use wild birds as natural hosts and reservoirs (Figure 4.2), bird surveillance is an important tool in arbovirus prevention and control programs (Figure 4.3). During the WNV outbreak

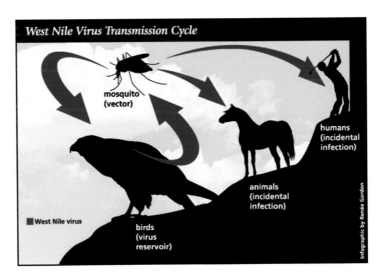

Figure 4.2 West Nile Virus transmission cycle.

Source: Photo courtesy U.S. Food and Drug Administration

Figure 4.3 Wild bird surveillance using a mist net.

Source: Photo courtesy Dr. Gail Moraru, used with permission

in New York City in 1999, crows began dying 6 weeks before the first human case.[7] Had this avian outbreak been recognized and investigated further (and mosquito control implemented), it is possible that the number of human cases might have been lower. The optimal bird species to be chosen for serologic surveillance in each geographic area should be determined by serosurveys. The best sentinels for serologic surveillance are those species in which infection is rarely fatal. Avian serology (other than hemagglutination inhibition and neutralization tests) requires a supply of species-specific antiserum. The responsibility for developing and distributing these reagents is generally shared by the CDC, USDA, and U.S. Geological Survey (USGS). Serum can be tested for antibodies to various agents. For polymerase chain reaction (PCR) or other methods to detect viral antigen in birds, the best sentinels may be those that are readily susceptible to viral infection (even those dying from the infection). Necropsy tissues from sick/dead birds can be studied by gross and histopathology, and tested by real-time (RT) PCR, virus isolation, and immunohistochemistry. There are some disadvantages to wild bird surveillance for arboviruses.

Disadvantages of Wild Bird Surveillance

1. Movement of free-ranging wild birds makes it impossible to know exactly where the infection was acquired.
2. Free-ranging birds must be live-trapped for serum collection, and state/federal permits are required for that.
3. State and federal bird capture permits require careful determination of species identity, sex, and age (i.e., someone needs to be trained in wild bird identification).
4. Venipuncture of small wild birds is technically difficult and can lead to bird mortality.
5. It is generally not feasible to serially bleed individual free-ranging birds because of low recapture rates (although banding can be useful).
6. Serologic testing may require species-specific antiserum.
7. If local laboratories begin conducting avian serology, large volumes of reagents will be required and there may be quality control/assurance problems.

Surveillance Using Nonhuman Mammals

In the United States, the USDA Animal and Plant Health Inspection Service, Veterinary Services (APHIS-VS) group conducts generalized passive surveillance and targeted active surveillance to detect zoonotic diseases among livestock.[8] While these efforts have traditionally been aimed at brucellosis and bovine tuberculosis, they have been expanded in recent years to include a wide range of maladies, including bovine spongiform encephalopathy, avian influenza, and WNV. Horses may be good sentinel animals for arboviral disease. For example, in WNV and EEE outbreaks, horse cases often precede human cases, and these data can be used as the basis to increase local mosquito control efforts.[9] However, the use of horse cases as a monitoring system is biased by the level of horse vaccinations in a given area. The EWT horse vaccine (eastern, western, and tetanus) (Figure 4.4) is very commonly used in many areas of the country, resulting in a large majority of horses being vaccinated. Veterinarians and veterinary service societies/agencies are essential partners in surveillance activities involving horses with neurologic disease. Serum and cerebral spinal fluid (CSF) may be collected for antibody testing, and necropsy tissues for gross pathology, histopathology, PCR, virus isolation, and immunohistochemistry.

Dogs and cats may sometimes be indicators of local disease transmission. Blood or tissue samples from pet dogs and cats with neurologic disease can be taken during clinic visits or upon necropsy.

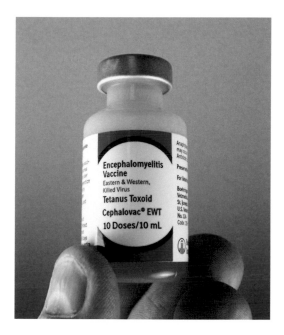

Figure 4.4 EWT (eastern, western, and tetanus) vaccine commonly used for horses.

Wild animals, even road-killed specimens, can be important sources of disease agent surveillance and ectoparasites. If animals are collected when freshly killed, blood, tissues, and ectoparasites such as lice and ticks can be removed and tested for disease agents. A lice comb can be used to comb the fur for ectoparasites (Figure 4.5).

Human Case Surveillance

Using humans as sentinel animals should be a last resort in public health programs and should not be used as an active method for detecting arbovirus activity, except possibly in states where (1) arbovirus activity is considered to be of very low likelihood or (2) resources to support avian-based or mosquito-based arbovirus surveillance are unavailable. Passive surveillance such as physician reporting is a crucial element in detecting emerging zoonotic diseases. During the 1999 WNV outbreak in New York City, by the time the health department received the first call from an infectious disease physician at the hospital in northern Queens, there were 15 similar cases in city hospitals that had not yet been reported.[7] Clinical syndromes to monitor in human surveillance programs include encephalitis cases (obviously this is the highest priority), milder illnesses

Figure 4.5 Wild animals may be important sentinels for disease agents.

such as aseptic meningitis and Guillain-Barré syndrome, or various other unexplained fever-with-rash illnesses. Specimens for collection and analysis include CSF, serum, and tissues. As early as the first few days of illness, IgM antibody to various encephalitis viruses can be demonstrated in CSF by antibody-capture enzyme-linked immunosorbent assay (ELISA). Virus may also be isolated (in appropriate biosafety laboratories only), or detected by RT-PCR in acute-phase CSF samples. Paired acute-phase (collected as early as possible after onset of illness) and convalescent-phase (collected a week or more after clinical onset) serum specimens are useful for demonstration of seroconversion to WNV and other arboviruses by ELISA or neutralization tests. Although tests of a single acute-phase serum specimen (IgM) can provide evidence of recent arboviral virus infection, a negative acute-phase specimen is inadequate for ruling out an infection. Paired samples are of critical importance in confirming infection. The CDC can distribute human WNV and other encephalitis virus antibody-positive control serum to public health laboratories for use in serologic testing. As for tissue samples, when arboviral encephalitis is suspected in a patient who dies, tissues (especially brain samples, including various regions of the cortex, midbrain, and brainstem) and heart blood can be submitted to the CDC or even specialized private laboratories for analysis. Individual tissue specimens should be divided, with half frozen at—70°C for later studies, and the other half placed in formalin. Tissues may be analyzed in a variety of ways—gross pathology, histopathology, RT-PCR tests, virus isolation, and immunohistochemistry.

Mosquito Surveillance

Mosquito surveillance, along with bird-based surveillance, should be the mainstay of most state surveillance programs for arboviruses, including WNV, EEE, LaCrosse encephalitis (LAC), and other viruses (see the next section for more detail). Adult mosquitoes can be collected using a number of traps, sorted by species, pooled, and tested for virus infection by conventional (endpoint) and real-time PCR or arbovirus detection assays such as RAMP®, VecTest®,[10] or VecTOR Test® (Figure 4.6). Blood meal identification can be conducted to determine principal vertebrate hosts of mosquito species.

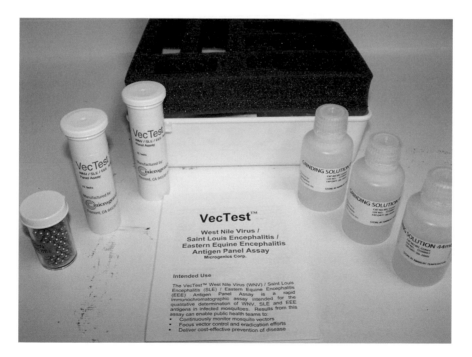

Figure 4.6 VecTest® and related products may be used for testing mosquitoes for arboviruses.

Advantages of Mosquito Surveillance

1. Provides the earliest and most definitive evidence of transmission in an area. Positive mosquito pools (from gravid traps) usually precede human cases by several weeks.[11]
2. Provides information on potential mosquito vector species in the area.
3. Provides an estimate of vector species abundance.
4. Provides information on virus infection rates in different mosquito species.
5. Provides information on relative risk to humans and animals.
6. Provides baseline and outbreak data that may be used to guide emergency control operations.
7. Allows evaluation of mosquito control methods and interventions.

Disadvantages of Mosquito Surveillance

1. Labor-intensive and expensive.
2. Substantial expertise is required for collecting, handling, sorting, species identification, processing, and testing.

Mosquito Surveillance

Mosquito surveillance should be a routine part of any arboviral prevention/control program. A good surveillance program can provide two types of information: (1) a list of the local mosquitoes (including distribution and population size estimates) and (2) the effectiveness of the control strategies being used. Routine surveillance can keep control personnel informed about locations of major breeding areas, helping to identify problem sites where control should be concentrated. Carefully interpreted survey data can provide vital information. For instance, large numbers of *Culex* egg rafts around the edge of ditches or *Aedes* eggs on oviposition strips are indicators that these breeding sites need to be watched closely the next few days. Treatment should be timed to catch the heavy crop of resulting larvae during the period of their life cycle when they are active feeders. Heavy adult catches in light traps stationed near treated areas may indicate that an important breeding site has been overlooked in the survey or that mosquitoes are migrating in from other areas, depending upon the species captured.

Mosquito Surveillance Networks

Most state health departments have an entomologist available to confirm field identifications (if not, see Chapter 8). Many municipal and county mosquito control agencies do not have the facilities or personnel available to identify mosquitoes. Therefore, local mosquito control agencies can collect mosquitoes using methods described below and send samples to their health department for identification, or be trained in mosquito identification by the public health entomologist to process their own samples. The mosquito sample can be identified and results sent back to the local mosquito control agency for an appropriate response. Mosquito control technicians or directors can consult the public health entomologist to determine what control tactics should be employed to control the problem species. Eventually, however, these control decisions will become second nature to local mosquito control personnel.

Establishment of fixed light traps (such as New Jersey light traps discussed later) can provide mosquito control personnel with valuable information about adult mosquito populations. Mosquito control agencies often station permanent light traps in the backyards of retired people living in mosquito-prone areas throughout the district. These retired individuals sometimes help with the surveillance process, maintaining the traps, collecting the mosquitoes after each sampling, and mailing the samples to the mosquito control agency. Operational expenses can be reduced by locating light traps at fire stations, city or county barns, water tanks, or other public facilities where they can be easily operated and maintained.

As an alternative, mosquito surveillance programs and larviciding activities offer excellent summer employment opportunities for local students. Students can provide a mosquito control agency with an economical, seasonally effective labor force. This program can also allow students to acquire experience in an applied aspect of biology, ecology, and public health science. Yearly training programs in mosquito control and surveillance techniques can be provided (with appropriate advance notice) by the local mosquito control agency or the health department.

Mosquito Egg Surveys

Oviposition jars or cups are useful tools for collecting information on container breeding mosquitoes, such as the Asian tiger mosquito (*Aedes albopictus*), the yellow fever mosquito (*Aedes aegypti*), and the tree hole mosquito (*Aedes triseriatus*). Counting eggs collected from an ovitrap will give a good indication of the number of *Aedes* larvae that will hatch in an area following the next rain. Eggs of some species can be quickly identified to species under a microscope (Figure 4.7). Also, eggs can be hatched in the lab and the mosquitoes tested for such things as pesticide resistance.

The oviposition cup should be a black plastic cup or glass jar, or even an aluminum can that has had the top cut out and painted black (Figure 4.8). The oviposition cup is fitted with a strip of felt-covered paper or a thin piece of wood clipped to the side. Some people simply insert "mosquito egg paper" into the cup. The cup is filled about halfway up with water. Female mosquitoes are attracted to both the black cup and the water, and they will lay eggs on the rough surface of the strip just above the water

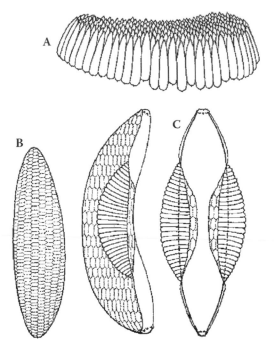

Figure 4.7 Eggs of mosquitoes: (A) Egg raft of *Culex* mosquitoes, (B) *Aedes* egg, and (C) *Anopheles* egg showing floats. *Source*: From the USDA Agriculture Handbook No. 173

Figure 4.8 Example of an oviposition jar/cup.

line rather than the smooth surface of the cup. A hole punched in the side of the cup about 2 in. from the rim will prevent water from flooding the eggs during heavy rains, thereby causing many of them to hatch. Some mosquito control districts attach an oviposition cup to both sides of a piece of white painted board. This arrangement reduces the chance of oviposition cups being turned over, and also, the contrast of black cups against the white boards seems to attract mosquitoes more readily.

Larval Surveys

The equipment required to conduct a larval surveillance program can be purchased in any hardware or home and garden store. A white plastic or a metal dipper is excellent for collecting water from artificial containers and small water bodies that are easy to reach (Figure 4.9). Larvae can then be gathered from the dipper with a medicine dropper and placed in a small jar, containing a little water, to be preserved later. Fancier, long-handled white graduated dippers can be bought from companies that supply mosquito control equipment (mosquito product vendors). These are useful for sampling ditches,

Figure 4.9 Dipping for mosquito larvae. *Source*: Photo copyright 2020 by Jerome Goddard, Ph.D.

Figure 4.10 Mosquito larvae in a jar.

margins of lakes and streams, and hard-to-reach areas. Kitchen strainers and fine-mesh aquarium nets are also good for collecting large numbers of larvae. The contents of the net or strainer can be washed into a white enamel pan or jar (Figure 4.10) and the large larvae removed with a medicine dropper. Large meat basters are ideal devices for getting samples from tree holes or artificial containers with restricted openings. Three- or 4-ft length sections of plastic tubing can also be used to siphon large amounts of water from tires and other similar breeding habitats. The end of the siphon can be placed in a strainer or large white pan to catch the larvae. A flat-weighted metal can with a string attached is an essential tool for collecting samples from storm drain ports protected by heavy metal gratings.

Potential mosquito breeding sites of all types should be sampled. Special attention should be paid to broken or leaking sewer pipes (towns and cities) (Figure 4.11) and malfunctioning wastewater disposal systems/septic tanks (rural areas), as these are particularly good breeding grounds for *Culex quinquefasciatus* and *C. pipiens* (Figure 4.12). Descriptive information should be written down for each larval sample collected. Accurate descriptions of habitats sampled, including those places where no mosquitoes are found, are equally important. By developing a good background on the type of local areas that breed mosquitoes, future routine surveys can be conducted efficiently, concentrating on known breeding sites. However, occasionally a thorough survey of all standing water areas should be conducted to ensure that previously unproductive areas have not (now) become mosquito breeding sites.

Estimates of population densities of larvae can be obtained by counting numbers of larvae per dip, using a standard size dipper (Figure 4.13 is an example of a sampling form). Three to ten dips should be taken and counted at each site. Number of dips counted and number of larvae in each

Figure 4.11 Leaking sewer pipe, providing good breeding site for mosquitoes.

Figure 4.12 Mosquitoes may breed in malfunctioning septic systems.

Mosquito Larval Surveillance Form

Date	Area or Sector No.	Anopheles			Culex			Aedes		Remarks (a)
		No. of Dips	Total Larvae and Pupae	Larvae/Dip and Pupae/Dip	No. of Dips	Total Larvae and Pupae	Larvae/Dip and Pupae/Dip	Type of Container	No. of Containers Positive for Breeding	

Adult Mosquito Surveillance Form

Date	Area or Sector No.	Species	Fixed Trap or Catching Station			Random Catching Spot or Station			Total				Density per Hour
			Time Spent	No. Collected M	F	Time Spent	No. Collected M	F	Time Spent	No. Collected M	F	Total	

a Remarks may be any climatological or environmental factors contributing to these particular samples.

Figure 4.13 Examples of forms for recording mosquito collection data.

dip should be recorded. Information on life stage of larvae and pupae can also be recorded. By noting numbers of larvae in each instar or size category (small, 0–5; medium, 5–15; large, 15+), number of pupae per dip, and water temperature, the investigator should be able to make an educated guess as to when mosquitoes will emerge and what types of control efforts should be used. In dengue surveillance programs, a container index is sometimes determined for larval *Aedes* mosquitoes in which the number of positive artificial containers is divided by the number of containers inspected × 100.

Generally, mosquito larvae develop faster at higher temperatures. Large numbers of pupae indicate that a large number of adults will emerge in a few days. Since pupae do not feed, use of *Bacillus thuringiensis israe-liensis* (BTI) or other products that must be eaten by mosquitoes will not control them. On the other hand, if most larvae are small, it may be 8–14 days before adults emerge, depending upon species and temperature. The investigator may then decide that an application of BTI is suitable. Large numbers of pupal skins floating on the surface are a sign that adult mosquitoes have recently emerged. An experienced investigator will also be able to determine the genus of many larvae based upon a few rather obvious characteristics. This knowledge will be useful in selecting the right larval control agent. For example, the bacterial product *Bacillus sphaericus* is more effective on *Culex* mosquitoes than *Anopheles* mosquitoes.

Mosquito larvae collected for identification should be handled carefully. When handled roughly, distinguishing bristles and other structures may become damaged, making identification difficult or impossible. Larvae that are to be preserved should be removed from the pan or dipper with a large-tipped medicine dropper, placed in a small jar containing water, labeled, and carried back to the office or lab.

Adult Surveys

Adult mosquito surveillance is a very important part of any mosquito program and is needed for disease prevention/control as well as justifying pesticide applications. Mosquito identifications can be done the traditional way (morphologically) with a microscope and published keys, or using newer molecular techniques. At least one research group has developed a repeatable and automatable DNA metabarcoding protocol that uses bioinformatic tools to identify mosquitoes to species with high accuracy from bulk samples.[12] Either way, records should be kept as to numbers of mosquitoes, date(s), collection methods, and the like (see Figure 4.13). Adult surveillance will provide information on effectiveness of the overall control program, but especially the larvicide program. However, presence of some adult mosquitoes does not mean larviciding efforts are not working. No program will be successful in totally eradicating mosquitoes. The objective is to control mosquito populations, keeping their numbers down to an acceptable level. Also, several species, such as the salt marsh mosquito (*Aedes sollicitans*) and the dark rice field mosquito (*Psorophora columbiae*), are capable of flying long distances and can move into an area from distant breeding sites.

Information that can be gained from routine adult mosquito surveillance includes:

1. Checklist of adult mosquito species in the local area
2. Estimate of adult mosquito population density and distribution
3. Indication of the presence of breeding sites that were overlooked
4. Identification of sites where larviciding efforts need to be increased
5. Source of adult female mosquitoes that can be used for arboviral surveys (testing by PCR, VecTOR test, or other commercial kits)

With the exception of mosquito landing counts, equipment needed to collect adult mosquitoes is generally more complicated and expensive than that required for collecting larvae. Adult mosquitoes are very fragile. They readily lose legs, scales, and wings when handled roughly, making identification difficult. Even with a microscope, adult mosquitoes that have not been processed properly can be hard to identify due to their fragile nature (Figure 4.14). The special collection methods and equipment described later are designed to assess adult mosquito populations with minimum damage to the specimens.

Landing Counts

A landing count or rate is an observation of the density of mosquitoes in an area with little or no concern for species diversity present. Landing counts, made using one to four people, are concerned with numbers of mosquitoes attempting to bite rather than the species composition[1] (Figure 4.15). A printed form is usually used to record this information. The form has a space to record start and finish times of the count, location at which it was taken, species collected, and some types of meteorological observations. If needed, specimens may be collected with an aspirator (Figure 4.16).

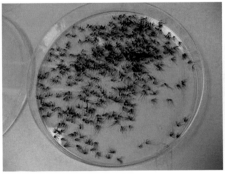

Figure 4.14 Identification of adult mosquitoes may be difficult.

Figure 4.15 Conducting landing counts for mosquitoes.

Source: Photo courtesy Dr. Wendy C. Varnado, used with permission

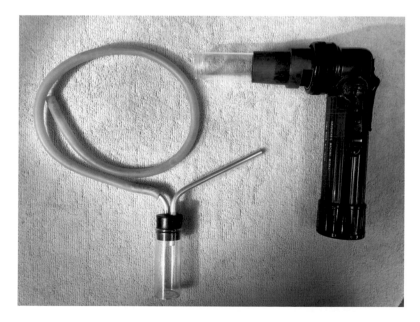

Figure 4.16 Two types of mosquito aspirators.

Source: Photo copyright 2020 by Jerome Goddard, Ph.D.

Most landing count surveys consist of biting attempts over a 5–10 min. period, and often mosquitoes are collected in small vials for identification to species. This information is kept on a special form. Landing rate data are used for many different purposes by mosquito control agencies, such as justification for adulticiding, effectiveness of adulticiding, vector potential of biting populations, light trap placement, and comparison of biting populations versus light trap data.

Daytime Resting Stations

Adult mosquitoes, especially *Anopheles*, can be found during the daytime resting in both natural and artificial shelters. These areas include houses, barns, sheds, privies, bridges, culverts, hollow trees, overhanging cliffs, and foliage. Counts of mosquitoes utilizing daytime resting shelters can give a good indication of population density. Mosquitoes found in these shelters can be easily collected with an aspirator. In areas where no resting shelters are found, an investigator may install an artificial shelter such as a wooden box, approximately 60 cm x 60 cm, so that these sites can be routinely sampled. Many mosquitoes that do not usually bite can be collected in this way.

Light Traps

Several types of light traps are commonly used in mosquito surveillance programs. The CDC light trap, developed by the Centers for Disease Control, is a widely used portable model (Figure 4.17). This light trap runs on a 6 V lantern battery; some versions use two D cell batteries. Mosquitoes are attracted to a small light at the top of the trap and are then sucked into a net at the bottom of the trap by a fan. The traps are usually set out and turned on at dusk and picked up at dawn. Timing devices can be installed on the traps so that they will only run during those hours of peak mosquito activity, conserving batteries. Only selected species of mosquitoes are attracted by light traps, and catches tend to be smaller during a full moon. Mosquito catches can be increased by hanging a container of dry ice or an octenol lure near the light trap.

The New Jersey light trap is a larger metal device, usually located at a permanent sampling station (Figure 4.18). This trap is often equipped with a timing device that turns it on during selected hours on certain days of the week. It works on the same general principle as the CDC light trap, except that it uses 110 AC power and mosquitoes are sucked into a paper cup inside a jar containing a killing agent such as a piece of DDVP (Vapona) pest strip. The paper cup prevents mosquitoes from coming into direct contact with the pest strip. Generally, New Jersey traps require little maintenance.

Figure 4.17 CDC light trap for mosquitoes baited with dry ice.

Figure 4.18 New Jersey light trap.

Oviposition or Gravid Traps

Oviposition traps or gravid traps are somewhat similar to oviposition cups in that they provide a black plastic container partially filled with water as an attractant (Figure 4.19). The gravid water is a mixture of either horse manure and water or fish oil and water that has had time to become putrid. Female mosquitoes visiting the trap to lay eggs are sucked into a net by a small fan motor like those used on many light traps. Oviposition traps are very selective for female *Culex* mosquitoes and are useful in WNV and SLE arboviral surveillance programs. The catch data are not comparable to light trap data.

Preserving Adult Mosquitoes

Adult mosquitoes should be handled very carefully to prevent them from losing scales, legs, or wings. Collections taken from a New Jersey light trap using a pesticide strip in a killing jar are usually dead in the perforated paper cup. These mosquitoes should be gently shaken from the cup into a small tissue-lined cardboard jewelry box, petri dish, or equivalent container. Mosquitoes should be arranged evenly over the tissue and one layer deep (Figure 4.20). A piece of tissue should be placed on top of the mosquitoes to prevent them from being shaken about, and the lid

Figure 4.19 Gravid trap.

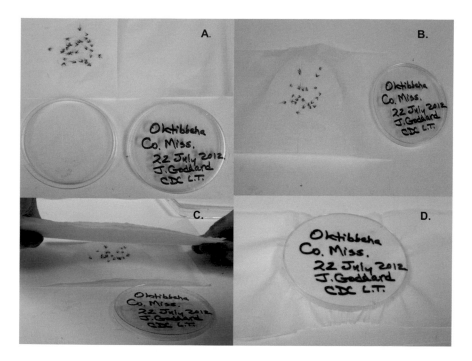

Figure 4.20 Steps to prepare mosquitoes for mailing: (A) sprinkle specimens on tissue paper, (B) place specimens and tissue in bottom half of petri dishes, (C) cover specimens with additional paper, and (D) replace top cover of petri dish.

replaced. A label containing necessary sample information can be placed on top of the last layer of tissue and the lid secured on the box with a rubber band. Only one sample should be placed per box or petri dish. If many mosquitoes are captured in a single trap, extra boxes or dishes can be used to hold the sample and these boxes bound together with a rubber band.

Mosquitoes can be removed from net bags of light traps with an aspirator (Figure 4.16). An alternative is to remove the net bag, tie the top, and place it in an airtight container such as an ice chest with an open bottle of chloroform (Note: Use with great caution with chloroform). Ice or frozen reusable ice packets can be used to "freeze" adult mosquitoes in an ice chest. This process takes longer than the chloroform method, and some mosquitoes may revive after being removed from the cold ice chest. Killed mosquitoes can be shaken from net bags onto a piece of paper or into a small pan and then transferred to a small collection box or petri dish. Adult mosquitoes should not be manipulated by hand, if possible. When it is necessary to pick them up, use a pair of forceps, grabbing each mosquito gently by a group of legs or a wing.

Mosquitoes can be mailed by placing petri dishes or collection boxes in a larger, sturdy container and filling loose spaces around the collection boxes with paper or Styrofoam peanuts. If the mosquito sample is to be tested for viruses, an ice pack should be included in the shipment. The lid should be carefully secured and the container marked "fragile." As an extra precaution, a layer of cotton can be placed on top of the upper tissue layer of each collection box and the lid replaced. This will help prevent mosquitoes from being shaken around inside the collection boxes. These steps may seem to be a little extreme; however, it is necessary to provide the identifying entomologist with good quality samples so that correct identifications can be made. This is especially important when adult collections are first being made from an area and a reference collection and species list is being compiled.

Tick Surveillance

Ticks "quest" for hosts by waiting on the tips of vegetation at variable heights (Figure 4.21). Only a portion of the ticks in a given population are questing at any one time as ticks periodically leave their questing sites to

Figure 4.21 Tick questing on vegetation. *Source*: Photo copyright 2017 by Jerome Goddard, Ph.D.

Figure 4.22 Tick surveillance using a drag cloth. *Source*: Photo courtesy Jerome Goddard II

move to the litter zone to rehydrate by active water sorption.[13] Since ticks quest for hosts, tick population densities can be estimated using tick traps baited with dry ice, dragging with a white flannel cloth, or walking surveys.[14–16] Drag cloth sampling (Figure 4.22), which is relatively inexpensive and easy to standardize by distance or duration,[17] has been shown to catch twice as many adult lone star ticks as walking surveys.[15] However, drag cloth sampling only measures the density of questing ticks and not the true population density. But rough population estimates can still be made since studies have shown that about 18–22% of adult ticks are questing at any one time.[18,19]

Habitat Mapping and Record Keeping

Habitat maps and records of mosquito and tick populations and pesticide applications are valuable sources of information for the public health entomologist. The following describes one method of mapping and keeping records of tick and mosquito activities. However, improvements and variations of this method certainly can be made. For mosquitoes, a habitat map should show all known water areas within a town, including artificial containers and floodwater areas. In the beginning of the larvicide program, all known water areas can be marked with a GPS unit or recorded upon photocopied quarter mile quadrants of a town street map or aerial photograph. For ticks, the habitat map should include areas where high tick numbers have been recorded.

The best way to conduct a habitat survey is by foot, inspecting each site for evidence of mosquito breeding or high tick populations. This insures a thorough inspection and allows the inspector to become familiar with the area. Tick and mosquito positive and negative sites should be recorded. Positive sites should be distinguished from negative sites by placing an asterisk (*) next to those sites where specimens were found. Water sites for mosquito breeding can be recorded by type, using a numerical code. Later, when making routine larviciding rounds, the technician can then quickly determine locations and types of water habitats in an area at a glance. He or she will then know which of these sites were positive during the initial survey.

Habitat maps can be verified during the first few applications of pesticides. Newly discovered sites should be added. Locations of ditches and storm drains, and tree lines can also be checked. When the field technician feels comfortable with the accuracy of the maps, photocopies can be made and a master copy kept on file. During each pesticide application trip, a set of these copies can be used as field maps. Notations concerning the day's activities can be recorded on each quadrant as the larvicide technician visits each site. Maps showing each week's activities can be kept on file for future reference on tick and mosquito population trends.

Some people laminate field and quadrant maps. During routine visits to these sites, the technician can make notations on the maps with a grease pencil. These maps should be kept in a metal clipboard with a cover in order to prevent smudging the grease pencil. Information should be transferred from laminated maps to regular copies or computer spreadsheets and filed for future reference. NOTE: much of present-day habitat marking is done with computer software such as Google Maps utilizing "drop a pin" technology.

Keeping records on each site can be useful. Knowing information such as previous pesticide treatments, past estimates of tick or mosquito numbers, life stages found, and when a site was wet or dry will allow the technician to predict when a particular site may become a problem. This information can be useful in other ways. Many vector-borne diseases occur in cycles. Knowing something about breeding trends of local tick and mosquito species over the past few years may indicate the likelihood of a disease outbreak. Any advance warning of a potential epidemic would allow public health entomologists to take precautions such as pesticide applications in an area more frequently.

Note

1. Asking personnel to do landing counts can lead to liability issues if someone contracts a mosquito-borne disease. Please consult your agency attorneys for legal advice in this matter.

References

1. Langmuir AD. Evolution of the concept of surveillance in the United States. *Proc R Soc Med.* 1971;64:9–12.
2. Herring ME. Where have all the vector control programs gone? *J Environ Health.* 2010;73:30–31.
3. CDC. CDC Guidelines for Arbovirus Surveillance Programs in the United States. In: Division of Vector-borne Infectious Diseases, Ft. Collins, CO; 1993:85 pp. www.cdc.gov/ncidod/dvbid/arbor/arboguid.htm.
4. CDC. Epidemic/epizootic West Nile Virus in the United States: guidelines for surveillance, prevention, and control, 3rd edition. In: Centers for Disease Control, Division of Vector-borne Infectious Diseases, Ft. Collins, CO; 2003:80 pp.
5. Tilak R. Entomology in public health practice. In: Bhalwar R, ed. *Textbook on Public Health and Community Medicine.* Pune, India: Department of Community Medicine, Armed Forces Medical College, in association with WHO/India; 2009:903–975.
6. Eldridge BF. Strategies for surveillance, prevention, and control of arboviral diseases in Western North America. *Am J Trop Med Hyg.* 1987;37 Suppl.:77S–86S.
7. Layton MC. Challenges of vector-borne disease surveillance from the local perspective: West Nile virus experience. In: Burroughs T, Knobler S, Lederberg J, eds. *The Emergence of Zoonotic Diseases.* Washington, DC: National Academy Press; 2002:86–90.
8. Crom RL. Veterinary surveillance for zoonotic diseases in the United States. In: Burroughs T, Knobler S, Lederberg J, eds. *The Emergence of Zoonotic Diseases.* Washington, DC: National Academy Press; 2002:90–96.
9. Morris CD. Eastern equine ecephalitis. In: Monath TP, ed. *The Arboviruses: Epidemiology and Ecology.* Boca Raton, FL: CRC Press; 1988:1–20.
10. Burkhalter KL, Lindsay R, Anderson R, Dibernardo A, Fong W, Nasci RS. Evaluation of commercial assays for detecting West Nile virus antigen. *J Am Mosq Control Assoc.* 2006;22(1):64–69.
11. Varnado WC, Goddard J. Use of the VectorTest for advanced warning of human West Nile cases in Mississippi. *J Environ Health.* 2016;79:20–24.
12. Mechai S, Bilodeau G, Lung O, et al. Mosquito identification from bulk samples using DNA metabarcodong: a protocol to support mosquito-borne disease surveillance in Canada. *J Med Entomol.* 2021;doi: 10.1093/jme/tjab046.
13. Needham G, Teel P. Water balance by ticks between bloodmeals. In: Sauer JR, Hair JA, eds. *Morphology, Physiology, and Behavioral Biology of Ticks.* West Sussex, UK: Chichester, Horwood, and Chichester; 1986:100–151.
14. Falco RC, Fish D. Potential for exposure to tick bites in recreational parks in a Lyme disease endemic area. *Am J Public Health.* 1989;79:12–15.

15. Schulze T, Jordan RA, Hung RW. Biases associated with several sampling methods used to estimate abundance of *Ixodes scapularis* and *Amblyomma americanum*. *J Med Entomol*. 1997;34:615–623.
16. Solberg VB, Neidhardt K, Sardelis MR, Hildebrandt C, Hoffman FJ, Boobar LR. Quantitative evaluation of sampling methods for *Ixodes scpaularis* and *Amblyomma americanum*. *J Med Entomol*. 1992;29:451–456.
17. Goddard J. Notes on the seasonal activity and relative abundance of adult black legged ticks, *Ixodes scapularis*, in Mississippi. *Entomol News*. 1986;97:52–53.
18. Goddard J. Proportion of adult lone star ticks (*Amblyomma americanum*) questing in a tick population. *J Mississippi Acad Sci*. 2009;54:206–209.
19. Huang M, Jones AM, Sabet A, et al. Questing behavior of *Amblyomma americanum* in a laboratory setting. *Sys Appl Acarol*. 2021;(in press).

chapter five

Regulatory, Political, and Legal Challenges

Helper versus Enforcer

Public health entomologists are public servants, with salaries paid with state or federal money. For this reason, they are obligated to help citizens with arthropod issues relating to public health. As described in earlier chapters, this can include identifying arthropod specimens submitted by the public or medical community, or on-site visits to schools, hospitals, or parks to investigate pest or vector problems. Generally, this is the helper role that is looked upon favorably by the public. In such situations, people might say something like, "I called the department of health about my problem and they offered all sorts of helpful advice and recommendations. I was very pleased with the way it all worked out." The other side of the coin is when the public health entomologist (PHE) must be a regulator, helping enforce state vector control or general sanitation rules and regulations. This kind of activity is necessary, but no fun for any of the parties involved. In such cases, the public may be angry and frustrated because the PHE made them clean up a breeding site or otherwise abate a nuisance. Perhaps the health department carried someone to court to force compliance with vector control rules. In that case, people might say something like, "$#&@#! Who owns this property anyway? Me or the health department? Can't I do what I want on my own property?" Complicating all this is the fact that politics may enter into enforcement decisions, or lead to reversals of previous decisions. To be an effective PHE requires knowledge of local and state political realities and an ability to use finesse to bring about enforcement with the least amount of political influence or damage.

Pests and Nuisances

Pests are generally defined as plants or animals, insects, fish, or other entities not under human control, which can be offensive or interfere with the comfortable enjoyment of life. They may be involved in mechanical transmission of disease agents (Figure 5.1). Health nuisances are what the name implies to both laypeople and the local or state health officer, and

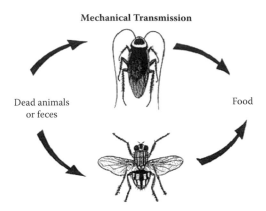

Figure 5.1 Mechanical transmission of disease germs by arthropods.

these mostly include breeding/growth sites of pests. For example, under this definition, any water source that is a breeding place for pests may be considered a nuisance (or at least a nuisance generator). Sometimes, these nuisances are not of significant public health importance at the current time (i.e., nobody is falling dead at the moment), but a health officer may be sufficiently concerned to seek a solution or abatement. From a public health standpoint, the chief nuisances are roughly grouped as follows: conditions favoring mosquito breeding (tire piles, ditches, swamps, etc.) (Figure 5.2), rat harborage (dumps, dilapidated buildings), fly breeding (manure piles, garbage) (Figure 5.3), and improper sewage disposal (failing or faulty on-site wastewater systems) (see Figure 4.12 in the previous chapter). However, from the standpoint of the public, anything offensive to the senses constitutes a nuisance and should be the basis of legislation. Unfortunately, they think this means odors. Note: Stench and odors may possibly be listed as a nuisance in some states. Therefore, a good definition of a nuisance is the use of one's property in such a way as to breed vermin, injure the rights of others, or inflict damages. Abatement of nuisances may be accomplished by regulatory action under statutory powers, injunction, criminal prosecution, or (private) civil suits for damages.

SOME PUBLIC HEALTH NUISANCES

1. Conditions favoring mosquito breeding (standing water)
2. Rodent harborages (dumps or dilapidated buildings)
3. Sites conducive to fly breeding (manure, garbage)
4. Improper sewage disposal (broken pipes, or failing wastewater systems)

Figure 5.2 Tire piles can be considered health nuisances due to their mosquito breeding potential.

Source: Photo courtesy Dr. Wendy C. Varnado

Figure 5.3 Poorly managed garbage can be a public health nuisance.

Public Health Emergencies

As discussed briefly in Chapter 1, the governor and state health officer in each of the United States are granted powers by state law(s) to respond to emergency situations. The actual definitions of "emergency" or "disaster" vary widely, and in some states remain undefined. In Indiana, for example, the Emergency Management and Disaster Law (Indiana Code §10–14–3) defines an emergency as the "occurrence or imminent threat of widespread or severe damage, injury, or loss of life or property resulting from any natural phenomenon or human act, including an epidemic and public health emergency."[1] Once an emergency is declared, designated officials can deploy the military, restrict freedom of movement, close or restrict businesses, and even suspend civil rights and liberties for a period of time. This type of executive authority was used widely throughout the United States and Europe during 2020 to help slow the spread of covid-19 (caused by the SARS-CoV-2 virus), resulting in severe economic losses and much public consternation. Hundreds of lawsuits against these government actions were filed. For example, taverns in Louisiana filed suit, challenging the governor's pandemic closure orders and mask mandate, alleging that the orders treated less fairly taverns versus similarly situated businesses, and thus deprived equal protection of laws and also, that the orders constituted an illegal commandeering of private property without just compensation.[2] In defending these various lawsuits, state officials claimed that the situation(s) was exigent, anticipated harm calamitous, and the harm could not be avoided through ordinary procedures. Most lawsuits questioning government covid-19 lockdowns were not successful, although those involving religious organizations were often overturned. In one such opinion, the U.S. Supreme Court stated, "The restrictions at issue here, by effectively barring many from attending religious services, strikes at the heart of the First Amendment's guarantee of religious liberty."[3]

Enabling Legislation

There is a hierarchy of law concerning public health, beginning with historical common law guiding kings, parliaments, and other governments in matters relating to the health, safety, and welfare of the public. Common law allows for basic sanitation and disease prevention to be a function of government. In fact, many people argue forcefully that government should be involved in the nation's healthcare.[4] In both the U.S. Constitution and the various state constitutions, there is recognition that states have police powers to enforce laws concerning the safety and welfare of their people. As far back as the mid-1800s, there was wide recognition for the need of "health police" to inspect, regulate, suppress, and prevent disease-producing circumstances.[5] This led to states passing legislation to provide for the safety and welfare of the population. For example, Mississippi Code § 41–3–15 allows for the establishment of a Mississippi Department of Health overseen by a board of health to have "general supervision of the health interests of the state," and, as another example, Georgia Code OCGA § 31–2–1 says, "The Department of Community Health shall safeguard and promote the health of the people of this state and is empowered to employ all legal means appropriate to that end." As for public health entomology in particular, Mississippi Code § 41–3–15 states, "The Mississippi State Board of Health shall have authority to establish programs to promote the public health in the area of food, vector control, and general sanitation." According to California law, the Department of Public Health may "maintain a program of vector biology and control including, but not limited to, the following: (a) providing consultation and assistance to local vector control agencies in developing and conducting programs for the prevention and control of vectors, (b) surveillance of vectors and vector-borne diseases, (c) coordinating and conducting emergency vector control, as required, (d) training and certifying government agency vector control technicians, and (e) disseminating information to the public regarding protection from vectors and vector-borne diseases" (Cal. HSC Code § 116110).

Few, if any, states have specific laws for every aspect of health, safety, and welfare. Instead, there is only broad legislation (like the previous Mississippi example) that enables the state health department to enact *rules* and *regulations* addressing these myriad issues. In most cases, rules and regulations are developed, then opened for public comment and input before being enacted by the department of health or, in some cases, an oversight group such as a board of health. Under this scenario, a health department rule or regulation carries the full weight of state law, even though it may not be specifically defined in state law. Again, using Mississippi as an example, Mississippi Code § 41–3–17 defines how rules and regulations can be adopted and their subsequent authority. Judges understand this and during enforcement issues, a rule or regulation is considered law. In my 20 years as a PHE, during enforcement proceedings, I never heard this questioned. Certainly, other ways were tried to get around enforcement, such as claiming the rule(s) is vague or overreaching or unevenly enforced, but never by questioning its validity or the underlying enabling legislation.

In contrast, *policies* may be developed by a health department that may help the staff administer and manage their rules and regulations. These are written and approved internally, and may help the agency prioritize resources, improve efficiency, and define exceptions to rules and regulations. Policies are often disparaged by critics as cumbersome and nothing but governmental red tape; however, they are useful to health department staff for interpreting and managing various aspects of regulations.

When Politics Interferes with Public Health

One of the most frustrating aspects of public health regulatory work is seeing a few people who seem to always get away with avoiding or breaking public health ordinances by complaining to politicians or invoking the power of a "friend" who has connections to a local judge. This is a reality of life, and often takes two forms: (1) complaining about possible personal health impacts of following an ordinance, such as those persons fearful of taking a required vaccination or refusing to allow mosquito spraying near their property, or (2) using political connections to avoid health department enforcement actions such as an order to fix a failing on-site wastewater system or clean up a food establishment.

Antivaccination and Antipesticide Fears

At least 25 vaccines have been developed or licensed for use against a variety of human diseases, and the Centers for Disease Control officially recommends routine vaccination to prevent 17 vaccine-preventable diseases that occur in infants, children, adolescents, or adults.[6] In addition, there are 9 other nonroutine vaccines recommended for people in certain travel or job situations. However, from the development of the first vaccine, there has been opposition to vaccination. The current resistance is mostly based on fears of possible side effects from vaccines such as autism, or the belief that vaccinations don't work at all. A smaller segment of the population refusing vaccinations distrusts government and may suspect racist or genocidal intentions. In the early 1980s, an inflammatory television program about the diphtheria-pertussis-tetanus (DPT) vaccine focused on a litany of unproven claims against it. During the ensuing public outcry, many countries dropped their mandatory DPT vaccination programs. Countries that dropped those programs subsequently suffered 10 to 100 times the pertussis (whooping cough) incidence than countries maintaining compulsory vaccination.[7] Little has changed in recent years, except perhaps the means and methods by which antivaccinationists get out their message. With the advent of the Internet and social media programs, powerful, even radical, groups have emerged that are able to sway public opinion and apply political pressure to allow exemptions from vaccinations. Interestingly, at one time, parents of unvaccinated children were mostly poor and without access to proper healthcare resources; now, however, these parents are highly educated and with middle to upper incomes. The trend is alarming. In some Marin County, California, schools (near San Francisco), more than half of kindergartners received "personal belief" exemptions from vaccinations during the time period 1999 to 2008.[8] And recently, some of the worst pertussis outbreaks in the last 50 years have occurred in California.[7] Persons requesting personal belief exemptions often fail to realize the serious threat posed by infectious diseases such as measles, polio, or pertussis. They somehow believe that a healthy, well-nourished human can successfully fight off infection with these diseases. One person interviewed in the Marin County newspaper article[8] cited previously said, "Vaccination is based on the medical fallacy that our bodies are stupid. The truth is that the body has nearly infinite capacity to protect itself against infection." The facts of history obviously speak otherwise.

In many ways, pesticide fears parallel vaccination fears, with similar social ramifications. Unfortunately, in our chemophobic society, people fail to appreciate the fact that research has shown that adult mosquito spraying is a cost-effective intervention for protection against mosquito-borne diseases.[9] Further, the U.S. Centers for Disease Control and the Environmental Protection Agency both recognize a legitimate and compelling need for the prudent use of space sprays, under certain circumstances, to control adult mosquitoes.[10] Guidelines provided by the Association of State and Territorial Health officers (ASTHO) also explain how certain "trigger" events may necessitate emergency mosquito spraying.[11] This is especially true during periods of mosquito-borne disease transmission or when source reduction and larval control have failed or are not feasible. However, fears about pesticide use are persistent in the United States, causing much controversy between those wanting mosquito control and those who do not. If, for example, spraying for mosquitoes in a municipality is suspended due to a few people insisting the products are making them ill, then that leaves the other (majority of) people susceptible to mosquito-borne diseases (Figure 5.4). Many local governmental leaders try to respect pesticide fears and sensitivities, posting their mosquito spraying schedules and even allowing "do not spray" forms to be submitted by the public that instructs the spray technician to cut off the spray machine within 500 or so feet on both sides of that property (Figure 5.5). Obviously, a no-spray zone cannot be unreasonably large. A woman once called me, insisting that I make her city stop mosquito spraying within a mile and a half of her house. She said if they didn't, she would die and then the city would be liable. I tried unsuccessfully

Figure 5.4 Some people are afraid of pesticide applications for mosquito control.
Source: Photo copyright 2020 by Jerome Goddard, Ph.D.

to explain that mosquito spray machines only produce a 300-foot swath (it's actually less than that), and that a mile-and-a-half no-spray zone was unreasonable and not fair to those in her neighborhood who did want mosquito spraying.

Under the County "no-spray list" law, Suffolk County Vector Control will maintain a registry of citizens who request a limited shut off of mosquito adulticiding (fogging). The law requires a "good faith effort" by Vector Control to shut off truck-mounted aerosol equipment within 150 feet of a registered property. **This registry will be rendered inactive if the Commissioner of Health declares a public health emergency for mosquito-borne disease. Therefore, this registry does not exempt properties from treatments for West Nile Virus, Eastern Equine Encephalitis or other mosquito-borne disease. In the event of mosquito-borne disease or aerial application, Vector Control will attempt to telephone members of the registry prior to treatment of their property.**

Please check one of the following options (either A or B) on this form to indicate your request; fill in the blank spaces and return the completed form to the Suffolk County Vector Control at the address at the top of this form. Your name, **a signature and all required information must be provided.** The request will take effect within 20 workdays of our receipt of this information.

☐ A. No mosquito adulticiding 150 feet in either direction in front of residence.

☐ B. Remove previously requested restrictions and resume normal operations.

This request is considered public information and the Division may notify neighbors as to why part of their neighborhood is not being treated if choice "A" is indicated. As part of its public notification process, a list of "no-spray" locations (without names) may be posted on the Vector Control Web site or made available by other means, and "no-spray" locations may be indicated on published and Internet treatment maps.

Name_____
          (Please Print or Type)

Signature _____ Date _____

Phone/Day (_____)_____ Phone/Evening (_____)_____

FAX (_____)_____ E-MAIL _____

Restriction address:	Mailing Address (if not same as restriction address):
(Street Number and Name)	(Street Number and Name)
City: _____State **NY** Zip Code: _____	City: _____ State ____ Zip Code: ____

Tax Map Number of restricted property (required – can be found on property tax bill or by contact with your Town):

District: _____ Section: _____ Block: _____ Lot: _____

REASON (optional)
_____

Applicants are required to file annually to maintain active status on this registry. The list is reviewed and updated annually, those who do not complete and return the form as required shall be removed from the registry.

Figure 5.5 Example of a Do Not Spray request form.

Using Politics to Avoid Enforcement

Enforcement sometimes entails taking a rule breaker before a local judge who is either sympathetic to the person, related to them (in rural areas), or friends with them. In that case, the health department regulatory official shouldn't be deterred, but instead strive to maintain professionalism, calmly and persistently explaining to the judge the public health importance of the enforcement issue in question. The enforcer should always remain calm and cool no matter the outcome. You win some and you lose some. Such is life. There will be another day, another chance to make the case before a judge. Eventually progress will be made, infractions remedied, and nuisances abated.

Establishing Collaborations with Public Officials

As odious as politics can be for most people, it is critical for public health administrators to establish healthy relationships with local, state, and federal politicians. For example, even at the local town or city level, it is the mayors, city council members, and/or aldermen[1] who make decisions about mosquito control. Although seemingly counter-intuitive, appeals from public health concerning implementation of proper mosquito control should be made directly to these local politicians, not to the actual people performing mosquito control activities. Public health entomology programs such as vector surveillance, determination of infection rates (lab studies), personal protection educational efforts, and vector control all rely heavily upon state and federal funding. If the public health agency neglects these political relationships and develops an "ivory tower" perception among citizens and their political representatives, many public health programs will cease, or at least be severely curtailed due to lack of funding. Once, when the author was a public health entomologist at the Mississippi Department of Health, he overheard health administrators discussing a list of representatives in the state legislature, and repeatedly, these administrators used words such as, "Yes, she's a friend of ours," or "No, that one's not on our side." Later, it became clear to the listener that this wasn't anything unethical or nefarious. They were simply reassessing their (public health positions) with members of the state legislature, and strategizing how to push through legislation favorable to public health goals and objectives. So, go for it! Learn to (legally) work the political system for public health.

Legal Aspects of Entomology and Pest Control

We live in an increasingly litigious society wherein just about every conceivable dispute in life or business is battled out in courts of law, sometimes even for low or trivial amounts of money awarded. It seems ridiculous that people can't work out their problems without taking every little thing to court. Unfortunately, entomologists and environmental health specialists (EHSs) are not immune to involvement in lawsuits filed by customers or citizens claiming neglect, harm, personal injury, emotional trauma, poisoning, or damage to property from insects. Particularly "hot" issues right now attracting lawsuits are bed bug infestations, damage from termites and other wood destroying insects, and pests found in food products. At the other end of this spectrum is the situation where entomologists or EHS personnel serve as expert witnesses in court cases. Under this scenario, the professional entomologist is serving as a consultant or testifying expert in a court case (Figure 5.6). They aren't being sued, but instead, are hired as technical experts in someone else's lawsuit, and many of the basic principles of record keeping, data integrity and transfer, and testifying are the same, no matter which side of the law you are on.

Figure 5.6 Dr. Goddard waiting outside courtroom, ready to serve as an expert witness.

Source: Photo copyright 2019 by Jerome Goddard, Ph.D.

Ways for a Public Health Entomologist to Avoid Lawsuits

Stay within your outlined professional "duties." Whether you work for a corporation, governmental entity, or private industry, one of the best ways to keep from getting sued is to understand your duties and perform them well (and *only* them). A corporate, public, or government employee is usually fairly protected from lawsuits if he/she remains within their prescribed duties. For example, if your job is to inspect restaurants, then *inspect* restaurants, following the policies and procedures outlined by the agency your work for. It sounds silly, but if you get sued, you can say something like, "Hey, I just followed the food regulation or IPM regulations adopted by my agency or health authority."

Stay within your expertise. Although similar to that discussed in the previous paragraph, there are important differences between duties and expertise. Never stray from your training or expertise, no matter how tempting. If, for example, you are a public health entomologist and a customer asks you about mold growth in their bathroom, it would be best to say something like, "I'm not an expert on indoor molds. The best thing for you to do is to ask someone else, maybe an industrial hygienist." As another example, if, during an investigation, you claim that the offending pest is a phorid fly, then you had better be able to demonstrate your fly identification abilities because this may be challenged in court. Attorneys for the other side will most certainly ask you questions like, "Mr. so-and-so, are you a board-certified entomologist?" Or, "Could you please tell us what kind of entomological training you have received." Or, "Mr so-and-so, how many phorid flies have you ever identified?" "And, what are the distinguishing characteristics of that particular fly family?"

Another common problem is when customers ask an entomologist a medical question or ask for a medical opinion. Nonphysician health department personnel should never offer any type of medical advice or examine a patient or any lesions/bites on their body. Invariably, at some point, the customer will become angry or disgruntled and claim that you offered a medical diagnosis, leading to harm or injury.

Maintain good notes and records of your work. Good record keeping is a must for both preventing and fighting lawsuits. All sorts of claims can be made by plaintiffs about you and your services, and if months or years have passed since the incident, written and digital records are critically important. Health department inspection reports or pest management "logs" or application records may be used to establish the mechanism of contamination in a single outbreak or to discern a pattern of neglect and sub-standard behavior. Records show what the "findings" were at the particular time of the incident, conditions, environmental data, and time and duration of your visit, and so on. For record keeping beyond standard inspection forms or pesticide application reports, this author suggests using a bound notebook (not loose-leaf) with consecutively numbered pages. This type of record book makes it difficult to claim that notes have been tampered with since the incident.

Good records may either help or hurt your case. Believe me, if the case goes to trial, your records will be prominently displayed on the wall in the courtroom for all to see. This can be quite embarrassing. I've personally seen lawsuits against pest control personnel won because the records revealed that the pest control technician was only at the site for <10 minutes (the argument being that it would be impossible to perform the necessary pest control applications in that short period of time). On the other hand, I've seen frivolous lawsuits defeated or thrown out because the records showed that the technician followed accepted inspection and treatment guidelines. Remember, people can claim or allege that you did all sorts of terrible things, but *proving* maleficence is another thing altogether. Records are an integral part of that proving/disproving process.

Read and follow pesticide labels (don't get off-label). In the same way that physicians are legally bound by prescription drug labels, EHSs and entomologists are mandated to use pesticides only according to their label directions. Using a product "off-label" is illegal and punishable by fines and even imprisonment. A sure-fire way to invite lawsuits is to use a product at a higher rate than prescribed on the label or in a site not listed on the label. Conversely, if someone sues you claiming injury from pesticide use, if your records show that you were within label rates and guidelines, the chances of a successful suit are diminished.

Cases That End Up in Court

By the very nature of their jobs, public health entomologists and EHSs at times may have their work records utilized in legal cases or courts of law. They may even be asked to give a deposition or testify in court. This could be anything from cases of public health nuisances, to mis-application of pesticides, to insect damage to foodstuffs or property. Knowing this, health inspectors should make every effort to make sure their records are reliable and will hold up when challenged in court.

Whether or not records or testimony is "admissible" is actually a big deal in legal matters, and is clarified in a later section. Attorneys and judges understand this concept, and much wrangling in court occurs over whether or not certain testimony or tidbits of information are admissible. Generally, statements made out of court, orally or in writing, are unreliable and inadmissible. However, even an out-of-court statement (especially business records) may sometimes be admissible when offered to "prove the matter asserted." The main thing is to make sure your records are accurate and reliable. The term "hearsay" is applied to testimony in a court proceeding where the witness does not have direct knowledge of the fact asserted, but knows it only from being told by someone else. Hearsay is often struck down in courts, for example, something like, "Jane told me that John Doe often doubles the rate of pesticides in the course of his work. I didn't see it myself, but that's what she told me." There are exceptions to the hearsay rule such as: (1) records of regularly conducted business activity may be admissible and (2) public records and reports may also be admissible if deemed reliable. For example, health inspector records, made during the regular course of one's duties, may be allowed by judges in court cases. As such, it makes sense for health inspectors to make efforts to strengthen the reliability of their records whenever possible. This might include things like supplementing written records with photographs whenever possible, and recording the author, date, and time of any notes placed in a file. This makes the data more reliable. The person with the most direct knowledge of the situation can then be found and questioned. In your notes, try to avoid saying things passively like, "the tenant was instructed about proper clean-up procedures and also was shown the proper method of waste disposal." *Who* instructed the tenant? *Who* showed them the proper method? Notes should be in an active voice, whenever possible.

There are two kinds of "professional" or paid experts in court cases—consulting experts and testifying experts. Consultants provide background and technical information to attorneys for a fee, but do not have to give a deposition or testify in court. They can be anonymous players in the background without fear of their opinions being attacked, cross-examined, or destroyed in a trial. They are merely offering up ideas and opinions to the attorneys who hired them. Consultants are important in court cases, especially in complex matters wherein almost no one understands the technical jargon. Attorneys for both sides need someone to clarify the issues and help them better prepare their case(s). Testifying experts, on the other hand, are hired by plaintiff or defense attorneys to testify in a deposition or in court to help their cause and win the case. In this situation, it is not just the facts of the case, but *how* the expert frames them, how he/she looks, and so on. It is all part of the package. Of course, testifying experts are supposed to be impartial and not be an advocate for the side they are working for (this is what I personally strive for). However, in reality, they sometimes want their side to win and may not offer information that hurts their cause. In fact, some testifying experts are downright dishonest and will say anything for money. These unethical professionals advertise in publications and trade journals as being "experts" in a wide variety of subject areas. One way to help expose these types of experts is to ask them what percent of their annual income is derived from professional testifying. Most real experts have another job or profession and testifying is only a part-time enterprise.

Scientific facts and expert testimony are subject to scrutiny. In the United States, federal and state courts now operate under "Daubert" rules of admissibility, based on a case in 1993, Daubert vs. Merrell Dow. In that case, the Supreme Court ordered a new standard for admissibility of scientific evidence now known as the Daubert test, which tries to ensure that so-called scientific evidence meets certain standards. Now, under the judge's new role under Daubert, expert reports or deposition testimonies need to address explicitly factors for reliability and relevancy.

Daubert Rules of Evidence Admissibility

The Reliability Factors in Daubert

- Has this scientific theory or technique been empirically tested?
- Has this scientific theory or technique been subjected to peer review and publication in scientific journals?
- What is the known or potential error rate in this theory or technique? Every scientific idea has Type I and Type II error rates, and these can be estimated with a fair amount of precision. There are known threats to validity and reliability in any tests (experimental and quasi-experimental) of a theory.
- What is the expert witness's qualifications and stature in the scientific community?
- Can this theory or technique and its results be explained with sufficient clarity and simplicity so that a court and jury can understand its plain meaning?

Expert witnesses will often be asked to give a deposition in a particular case and (further) may be required to testify in the actual trial, should the case go to trial. A deposition is part of the process of assembling evidence before the trial in a lawsuit. Depositions may be taken anywhere, but a court reporter is there to take the deposition, and the testimony is "sworn" or under oath. The "deponent" is the person being questioned, and may be either the plaintiff, the defendant, or various experts/witnesses. If you are ever asked to give a deposition, at the beginning you will be asked to offer all sorts of background information about who you are, your educational and technical background, and then, more importantly, *why* you are uniquely qualified to offer an opinion in the case. Be careful during depositions to only say what you *intend* to say and nothing extraneous. Attorneys may try to set you at ease to get you to answer questions freely—to offer up new or unintended information. This can come back to haunt you. Attorneys may ask you leading questions to arrive at (their) desired answers, or they may ask you several questions in rapid-fire fashion, leaving you little time to think through your responses. Don't be tricked. You have the right to slow down and think about each question before responding. You also have the right to ask for a question to be repeated or even rephrased. In one deposition I was giving, I asked for the same question to be rephrased three times. I wanted to make sure what I said was what *I wanted to say.* Also, attorneys from the opposing side may try to discredit or destroy you during cross-examination. They may offer up hypothetical scenarios, such as, "What would you say, Dr. Goddard, if I told you we have evidence directly contradicting what you're saying here today?" Remain calm and don't let it unsettle you. Tell them, "Fine, then bring it out. Let's see your so-called evidence." Chances are, they have no such evidence.

Entomologists, health inspectors, and other health department personnel need not be fearful of courts or legal proceedings. Keep a cool head, rely on the notes and records you made during your visits, and don't be afraid to say, "I don't know," or "I don't recall." It's better to say that than to piece together an answer that is sketchy, even faulty. Keep in mind that if you choose to be an expert witness (for compensation), it's an adversarial environment and you will need a tough skin to withstand the cross-examination and (sometimes) attack on your integrity and character.

Note

1. Official names and designations of local officials vary by state or country.

References

1. Haffajee R, Parmet WE, Mello MM. What is a public health emergency? *N Engl J Med.* 2014;371:986–988.
2. Anonymous. Lawsuits about state actions and policies in response to the coronavirus pandemic, 2020. In: Ballotpedia; 2020. https://ballotpedia.org/Lawsuits_about_state_actions_and_policies_in_response_to_the_coronavirus_(COVID-19)_pandemic,_2020 (accessed December 1, 2020).
3. Barnes R. Supreme court relieves religious organizations from some covid-related restrictions. In: Washington Post, November 26 issue; 2020. www.washingtonpost.com/politics/courts_law/supreme-court-relieves-religious-organizations-from-some-covid-related-restrictions/2020/11/26/305f0094-2fa6-11eb-860d-f7999599cbc2_story.html.
4. Hamowy R. *Government and Public Health in America.* London: Edward Elgar Publishing, Inc.; 2007.
5. Griscom JH. Discourse in 1855 before the New York Academy of Medicine. In: Rosenberg CE, ed. *Origins of Public Health in America.* New York: Arno Press; 1972:31–58.
6. CDC. Current recommendations on vaccines. In: CDC, MMWR. 2011;60(RR02):1–60.
7. Poland GA, Jacobson RM. The age-old struggle against the anti-vaccinationists. *N Engl J Med.* 2011;364:97–99.

8. Rogers R. Refusal to vaccinate puts kids at risk. In: The Marin Independent Journal, Marin County, CA. 2009;April 11, issue. www.marinij.com/ci_12125899?IADID=Search-www.marinij.com-www.marinij.com.

9. Luz PM, Vanni T, Medlock J, Paltiel AD, Galvani AP. Dengue vector control strategies in an urban setting: an economic modelling assessment. *Lancet.* 2011;377(9778):1673–1680.

10. CDC. Joint statement from the CDC and EPA concerning mosquito control. In: U.S. Centers for Disease Control and the Environmental Protection Agency; 2007. www.epa.gov/opp00001/health/mosquitoes/mosquitojoint.htm.

11. ASTHO. Before the swarm: guidelines for the emergency management of mosquito-borne disease outbreaks. In: Association of State and Territorial Health Officers (ASTHO), Arlington, VA, Publication PUB-0807001; 2008:22 pp.

chapter six

Public Health Entomology Preparedness

"Americans are at an increasing risk of vector-borne diseases, and the United States is not adequately prepared to respond to these threats."

U.S. Centers for Disease Control, September 2020

Disease Emergence

Public health entomologists must always be prepared to deal with vector-borne disease outbreaks. These may be old or new threats. New or at least newly described diseases, the so-called emerging diseases, continue to be described. Many of these emerging or reemerging infectious diseases are vector-borne, meaning they are transmitted by insects or other arthropods. The United States Centers for Disease Control recently issued a national framework for the prevention and control of human vector-borne diseases, including participating stakeholders[1] (Figure 6.1). Vector-borne diseases are an especially significant human health burden in tropical areas (Figure 6.2). Malaria—the number one vector-borne disease worldwide—continues to be a huge public health crisis in many areas (Figure 6.3). There are an estimated 230 million cases of malaria worldwide each year, with about a half million deaths.[2] Several factors are responsible for the continuing scourge of malaria: (1) massive environmental changes affecting *Anopheles* mosquito populations in endemic areas, (2) insecticide resistance in the vector mosquitoes, (3) drug resistance in the malaria parasite, (4) deforestation, (5) human population migrations, and (6) increased travel by nonimmune expatriates.[3,4] Since 1975, the mosquito-carried disease, dengue fever, has surfaced in huge outbreaks in more than 100 countries (Figure 6.4). Some experts estimate that there may be as many as 400 million cases of dengue each year,[5,6] and the disease is spreading, even here. Approximately 25 locally acquired cases of dengue occurred in the Florida Keys in 2009, and about 60 in 2010.[7,8] Dengue is called break-bone fever because the classic form is characterized by sudden onset of fever, frontal headache, retro-orbital pain, and severe myalgias. The more dangerous form of the

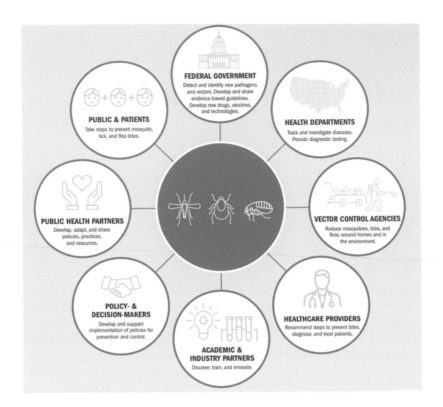

Figure 6.1 CDC national framework for surveillance and control of vector-borne diseases.

Source: Figure from the Centers for Disease Control

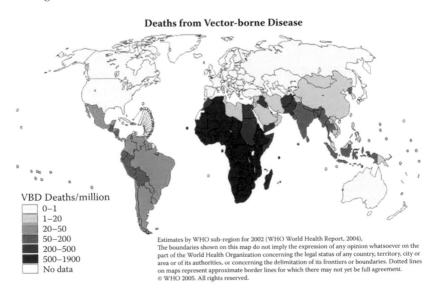

Figure 6.2 Deaths occurring from vector-borne diseases.

Source: Figure from the World Health Organization

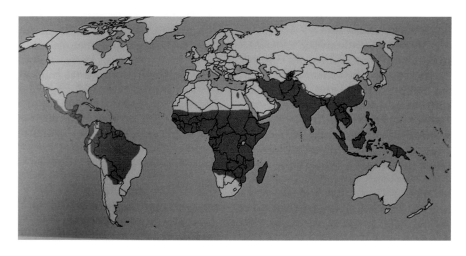

Figure 6.3 Distribution of Malaria.

Source: Figure from the Centers for Disease Control

Figure 6.4 Approximate distribution of dengue.

disease, dengue hemorrhagic fever (DHF) and dengue shock syndrome (DSS), with internal bleeding and shock, has been emerging over the last few decades, mainly affecting children under age 15 (Figure 6.5). DHF/ DSS can be a dramatic disease with the patient's condition deteriorating rapidly.

Figure 6.5 Increase in dengue hemorrhagic fever in the Americas.

Source: Figure from the Centers for Disease Control

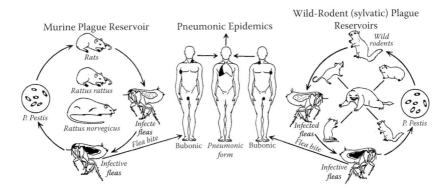

Figure 6.6 Plague life cycle, showing involvement of both fleas and rodents.

Source: Figure from the Centers for Disease Control

The flea-transmitted disease plague is also a constant threat (Figures 6.6 and 6.7). Since the early 1980s there has been a general trend toward increase in the annual number of cases to nearly 3,000 per year, and an increase in the proportion of cases reported in the African region.[9] In the United States, cases have traditionally occurred out west, but more recently, there has been an eastward movement in human cases toward the 100th meridian.[10] In 2015, 16 cases of plague were reported to the CDC from Arizona, California, Colorado, Georgia, Michigan, New Mexico, Oregon, and Utah,[11] but only one reported case in 2018.[12] The parasitic disease leishmaniasis, carried by sand flies, continues to be a problem as well. Visceral, cutaneous, and mucocutaneous forms of the disease occur. But

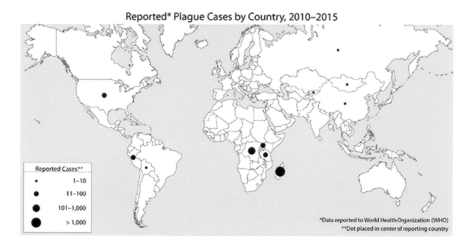

Figure 6.7 Worldwide distribution of plague.

Source: Figure from the Centers for Disease Control

perhaps the worst is mucocutaneous, wherein much of the face may erode away, leaving the patient hideously disfigured. The incidence of cutaneous leishmaniasis is on the rise in Central and South America because of road building, mining, oil exploration, deforestation, and establishment of communities adjacent to primary forest. Cutaneous leishmaniasis was a significant problem during the war in Iraq, with many soldiers returning with "Baghdad boil." Locally acquired cutaneous leishmaniasis does occur in the extreme southern parts of the United States, but is uncommon. Once thought to be absent from the United States, visceral leishmaniasis may now be established here. One article detailed the finding of more than 1,000 hunting dogs infected with the disease from 21 U.S. states and Ontario, Canada.[13] Lyme disease, unheard of in 1979, is now the number one tick-borne disease in the United States, with approximately 35,000 cases reported each year (some say as many as 300,000 cases per year).[12,14] In Europe, the number is easily hundreds of thousands of cases.[15] Rocky Mountain spotted fever and related diseases such as the newly described American Boutonneuse fever (*Rickettsia parkeri* infection) are rapidly increasing in numbers in the United States. These two and other related illnesses are now combined into one reporting category at the CDC—spotted fever group rickettsioses (SFGR).There were 5,544 cases of SFGR reported to the CDC in 2018.[12] Other tick-borne diseases such as babesiosis and ehrlichiosis are also emerging.[16,17] Several new *Babesia* species infecting humans have been found.[18] Likewise, there are at least three *Ehrlichia* species in the United States that produce spotted fever-like illnesses, one of them fairly recently recognized and called the *Ehrlichia muris*-like agent or *E. eauclairensis*.[17,19,20] Others will likely be found.

**FOUR WAYS VECTOR-BORNE DISEASES COULD BE
REINTRODUCED INTO THE UNITED STATES**

1. Through infectious vectors
2. Through infectious humans
3. Through infectious hosts
4. Through intentional release (bioterrorism)

It could be argued that at least some of this increase in vector-borne disease is due to increased recognition and reporting. Maybe we're just getting better at diagnosing these things. As opposed to older, serological diagnostic methods, specific disease recognition is certainly made easier by modern laboratory technologies such as polymerase chain reaction (PCR). However, changes in society such as population increases, ecological and environmental changes, and especially suburbanization (building homes in tracts of forested lands) are contributing to an increase in incidence of many of these vector-borne diseases.

Role of Climate Change in Vector-Borne Disease Outbreaks

Most scientists agree that the earth is warming, and since arthropods are ectothermic (combined with the fact that the extrinsic incubation period of pathogens is temperature dependent), a warmer climate should affect the status of various vector-borne diseases. For example, during the first two decades of the 21st century, invasive mosquitoes became established across Europe, with subsequent outbreaks of dengue fever and chikungunya virus. Further, tick-borne encephalitis and Crimean-Congo hemorrhagic fever changed their geographic distributions in response to warming temperatures.[21] The 2014 Intergovernmental Panel on Climate Change emphasized the significance of potential changes in vector-borne infectious diseases.[22] There is evidence that climate change helps spread vector-borne diseases to new geographic areas. For example, malaria has recently spread into highland regions of East Africa where it previously did not exist. This spread presumably occurred because of warmer and wetter weather, resulting in high rates of illness and death because the disease was introduced into a largely non-immune population.[23,24]

However, even though scientists agree that climate change will (likely) lead to an increase in incidence and scope of vector-borne diseases, it is unwise and an oversimplification to make assessments of the effect of changing temperatures alone on diseases. There are many factors impacting vector-borne diseases worldwide, many of which not only affect disease incidence directly, but also when in concert with other factors such as urbanization, land use changes, human and animal migration patterns, habitat fragmentation, and so on.

Prevention of Human Risks Associated with Travel and Arthropod Contact

International travel is a major problem for public health officials as it leads to spread of diseases and exotic pests. We are, in fact, now one global community sharing each other's diseases. A report by the Institute of Medicine has identified "increases in travel and commerce" as a major contributor to infectious disease emergence and reemergence.[25] Adventure travel and "eco-vacations" are a large and growing segment of the leisure travel industry, with growth of 10% per year since 1985.[26] People are bringing exotic pests and diseases back from these tropical locations, and health officials must at all times be prepared to identify, treat, and eradicate these threats.

Travel medicine issues are usually handled by health departments, private travel medicine clinics, or travel medicine departments at large medical institutions or medical schools. These health experts advise international travelers about the health risks associated with places they plan to visit and offer vaccinations or prophylactic treatments to help avoid illness while traveling. The standard reference for healthcare providers advising such travelers is the Yellow Book.[27] In addition, the CDC has a very useful webpage dedicated to traveler's health, which has health information for over 200 international destinations, diseases related to travel, and vaccinations.[28] Persons planning trips overseas, and especially to tropical countries, should contact their local health department for advice or make an appointment with a travel medicine expert at a local medical center.

Disaster Vector Control

The number of reported natural disasters worldwide is increasing.[29] Changes in global weather patterns and degrading of the environment mean that such calamities are increasing in frequency, complexity, scope, and destructive capacity. There are innate, built-in, natural protections against disasters, but that capacity is being weakened by destruction of forests and wetlands worldwide. In addition, climate change may increase the risk of storms, drought, and coastal flooding. Half of the human population lives within 50 miles of a sea coast or sea-navigable waterway, and 8 of the 10 largest cities in the world are located on the coast.[29] While natural disasters of many types can lead to a breakdown of civilization and disease control infrastructure, insect and vector problems are especially associated with floods and hurricanes (typhoons). Floods affect an average 520 million people each year.[29] Almost half the people killed in disasters in recent decades have been victims of floods. Hurricanes can bring excessively destructive winds and storm surges in their wake. For example, a typhoon accompanied by an exceptionally high storm surge swept over the coastal wetlands of Bangladesh in 1970, killing 300,000 people. In 2004, immediately following the catastrophic Indian Ocean tsunami, government officials from many nations met in Kobe, Japan, for the Second World Conference on Disaster Reduction. At that conference they adopted what is known as the Hyogo Declaration, a framework aimed at building the resilience of nations to disasters (see text box).

THE FIVE HYOGO COMMITMENTS (WORLD CONFERENCE ON DISASTER REDUCTION, DECEMBER 2004)

1. Make disaster reduction a priority.
2. Know the risks and take action.
3. Build understanding and awareness.
4. Reduce risk.
5. Be prepared and ready to act.

There are all sorts of insect and rodent vector problems resulting from disasters, and these pests may occur sequentially (Table 6.1). These problems ensue as a result of breakdown of basic sanitation services such as lack of clean water and appropriate sewage and garbage disposal, as well as new and abundant breeding areas resulting from the disaster. Primary pests related to disasters such as floods or hurricanes are mosquitoes, filth flies, and rodents. Mosquitoes breed in waters from floods or heavy rains, while filth flies and rodents breed in decaying organic matter. It is noteworthy that in a flood event, such as the great Mississippi River flood of 2011, mosquito problems are not apparent until *after* the flood itself. As floodwaters recede, the first problem encountered is huge numbers of floodwater mosquitoes in the genera *Aedes* and *Psorophora*. Members of these genera are not principal vectors of disease, but they are aggressive biters and numbers can be large enough to hamper outdoor activity. The second problem that may then follow the flooding event is increased numbers of disease vectors belonging to the genus *Culex*. These mosquitoes develop in polluted standing waters and sometimes artificial containers

Table 6.1 Likely Vector Control Issues Sequence after Disasters.

Within the first few days	7–14 days	15–30 days	31–180 days
Nothing	Floodwater mosquitoes such as *Aedes* and *Psorophora* Nuisance only—not vectors of disease	*Culex* mosquitoes breeding in polluted standing water and sometimes artificial containers lying around	Rodent breeding in debris piles
	Blow flies and house flies breeding in spoiled food, dead animal carcasses, and debris piles	Blow flies and house flies breeding in spoiled food, dead animal carcasses, and debris piles	Other nuisance pests breeding in debris such as brown widow spiders, snakes, etc.
	Possibility of diarrhea, dysentery, conjunctivitis, and other mechanical transmission of disease agents	Possibility of diarrhea, dysentery, conjunctivitis, and other mechanical transmission of disease agents	
		Rodents breeding in debris piles	
		Possibility of rat bites, leptospirosis, and mechanical transmission of disease agents	

lying around in the floodplain. Other common mosquito breeding sites after disasters include damaged, malfunctioning, or abandoned swimming pools, trash and debris holding water, and sewage arising from damaged wastewater systems. As for flies and rodents, debris piles after disasters may become huge pest-producing heaps that remain for months. Typically after a hurricane, debris is pushed to the sides of roads with bulldozers or other heavy equipment awaiting pickup by contractors hired by the local or federal emergency management agency (Figure 6.8). Compounding the problem is the fact that citizens returning home begin cleaning out their houses and refrigerators of spoiled food, all of which ends up in the debris pile(s). This is perfect breeding grounds for flies and rodents.

Figure 6.8 Debris pile on side of road after Hurricane Katrina.

Source: Photo courtesy Jerome Goddard, Ph.D.

Pests Involved

The major mosquito pests after natural disasters include the floodwater mosquitoes, and secondarily, the container-breeding species occurring in trash and debris left deposited after the storm.[30] In Europe, floodwater species such as *Aedes sticticus* and *Ae. vexans* may emerge in huge numbers after prolonged heavy rainfall or river inundation covering large expanses of land, including settlements.[31] In the Americas, these same two species, as well as *Psorophora* species such as *Ps. columbiae*, *Ps. ferox*, and *Ps. mathesoni*, are major pests after floods and storms. Nuisance biting by any of the above species may become unbearable at times. Container-breeding species in Europe include *Culex pipiens*, *Aedes albopictus* (an invasive species), and *Anopheles plumbeus* (a species that seems to be expanding its breeding habitats). In the United States, container-breeders include *Ae. aegypti*, *Ae. albopictus*, *Ae. triseriatus*, *Cx. pipiens*, and *Cx. quinquefasciatus*, among others.

Flies occurring in high numbers after disasters primarily belong to the insect families *Muscidae* (mostly house flies), *Calliphoridae* (blow flies), and *Sarcophagidae* (flesh flies), which are domestic nonbiting flies commonly seen in and around human dwellings. They are often collectively called filth flies, and may breed in various sorts of decaying organic matter after disasters. These flies do not bite, but are medically important in the mechanical transmission of disease agents from feces or dead animals to foodstuffs or food preparation areas. Throughout the world, these flies serve as carriers of organisms causing diseases such as typhoid, diarrhea, amoebic dysentery, cholera, giardiasis, pinworm, and tapeworm. They may spread these agents via their mouthparts, body hairs, or sticky pads of their feet, as well as through their vomit or feces. In addition, the egg-laying habits of these insects may lead to another human malady, myiasis, which is infestation of people by the maggots of flies. Filth flies often lay eggs on dead animals or decaying organic matter; therefore it is understandable that they might occasionally mistake a neglected wound on a person as a "dead thing" and thus lay eggs on it. This type of myiasis is opportunistic (sometimes referred to as facultative), and rarely, if ever, leads to infestation of healthy tissues.

Rodent pests occurring after disasters are primarily the commensal rodents, which are the brown rat (also known as the Norway rat), the roof rat, and the house mouse (Figure 6.9). Other wild rodents may increase in numbers as well after disasters, but commensal rodents are the main pests to contend with and may be involved in mechanical disease transmission (see Chapter 19).

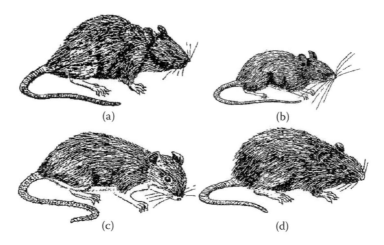

Figure 6.9 Various rodent species: (A) domestic rat, (B) house mouse, (C) rice rat, and (D) cotton rat.

Source: Figure from the Centers for Disease Control

Pest Control Options

Pest control after disasters relies heavily on sanitation (cleaning up or abating the breeding conditions), but also includes use of residual pesticides. Rodent control involves careful placement of rodenticides (please check local and national laws concerning rodenticide use outdoors) and physical trapping (Figure 6.10). For mosquito control, larviciding with chemical or biochemical insecticides works well to reduce breeding, but adulticiding with truck-mounted ULV spray machines is often necessary. Filth fly control involves local space spraying where needed, and application of fly baits where allowed by the insecticide label (Figure 6.11). In extreme cases, application of insecticides by airplane may become necessary for filth flies and mosquitoes (see the following examples). It should be noted that because of the filtration effect of dense foliage or broken down (open) housing, sometimes the insecticide dosages may need to be increased as much as twofold to achieve satisfactory mosquito control.[32] However, in many cases, 2× dosages are prohibited by the insecticide label. While aerial spraying of pesticides may be the last resort, it is nonetheless the most effective for rapid reduction of pests and for breaking vector-borne disease transmission cycles (see graph of mosquito control results in Figure 6.28 and these references).[33,34] Even under the best conditions, repeated adulticide treatments may be necessary to lower mosquito populations sufficiently to stop disease transmission.[35]

Figure 6.10 Rodent control through baiting and trapping.

Source: Figure on right courtesy Centers for Disease Control

Figure 6.11 Dead flies after application of granular fly bait.

Role of FEMA after Natural Disasters

The mission of the Federal Emergency Management Agency (FEMA), housed in the Department of Homeland Security (DHS), is to protect the United States from all hazards and reduce loss of life and property. FEMA, in its various early forms, has a long history. The first legislative act of federal disaster relief in U.S. history followed a devastating fire in Portsmouth, New Hampshire in December 1802. The destruction of large areas of the city's seaport threatened commerce in our newly founded nation. Therefore, the U.S. Congress provided relief to affected Portsmouth merchants in 1803 by suspending bond payments for several months. In the 1970s, President Jimmy Carter signed Executive Order 12127 establishing FEMA. Shortly after, in signing Executive Order 12148 on July 20, 1979, President Carter expanded the agency's role to the dual mission of emergency management and civil defense. FEMA's authority was further defined and expanded by the Disaster Relief and Emergency Assistance Amendments of 1988, which amended the Disaster Relief Act of 1974 and renamed it the Robert T. Stafford Disaster Relief and Emergency Assistance Act. This Stafford Act provided clear direction for emergency management and established the current statutory framework for disaster response and recovery through presidential disaster declarations. The terrorist attacks of September 11, 2001 drew new focus to homeland security and emergency management issues, leading to major statute and policy changes to reorganize the federal government. In 2002, President W. Bush signed the Homeland Security Act, leading to the creation of the U.S. Department of Homeland Security (DHS). The department was created on March 1, 2003 which united FEMA and 21 other organizations. After Hurricane Katrina made landfall in Mississippi, causing large-scale devastation along the Gulf Coast and resulting in billions of dollars in losses to infrastructure and the economy, Congress passed the Post-Katrina Emergency Management Reform Act of 2006 which established FEMA as a distinct agency within the DHS. Federal capabilities were once again tested in 2012 when Hurricane Sandy affected the entire East Coast. In response, Congress passed the Sandy Recovery Improvement Act of 2013 which streamlined the recovery of public infrastructure and allowed Federally recognized tribes to directly request a Presidential declaration. More recently during 2017, the nation faced a historic Atlantic hurricane season along with extreme wildfire disasters. This unprecedented and rapid succession of disasters transformed emergency management and focused efforts to build a culture of preparedness, ready the nation for catastrophic disasters, and reduce FEMA's complexity. Congress provided the agency with expanded authorities to further these goals by enacting the Disaster Recovery Reform Act of 2018. The legislation highlighted the federal government's commitment to increasing investments in mitigation and building the capabilities of state, local, tribal, and territorial partners.

FEMA'S EMERGENCY MANAGEMENT ROLE

1. Raise risk awareness, educate about risk reduction options, and take appropriate actions **before disasters**.
2. Alert, warn, and message affected communities, coordinate the Federal response, and apply/manage resources **during disasters**.
3. Coordinate Federal recovery efforts, provide resources, and apply insight as to future risks **after disasters**.

In responding to an event such as a hurricane or other disaster, FEMA Essential Support Functions (ESFs) are deployed to the affected area (Table 6.2). One such ESF commonly deployed after a hurricane is ESF-8, Public Health and Medical Services. FEMA may become involved

Table 6.2 Federal Emergency Support Functions (ESFs) and Their Areas of Responsibility.[1]

ESF designation	Area of responsibility	Primary lead agencies
ESF1	Transportation	Dept. of Transportation
ESF2	Communications	National Communications system
ESF3	Public works and engineering	Army Corps of Engineers
ESF4	Firefighting	Dept. of Agriculture/ Forest Service
ESF5	Information and Planning	Federal Emergency Management Agency
ESF6	Mass care, emergency assistance, temporary housing, and human services	Dept. of Homeland Security/American Red Cross
ESF7	Logistics	General Services Administration/FEMA
ESF8	Public health and medical services	Dept. of Health and Human Services
ESF9	Search and rescue	Federal Emergency Management Agency
ESF10	Oil and hazardous materials response	Environmental Protection Agency
ESF11	Agriculture and natural resources	Dept. of Agriculture/ Interior
ESF12	Energy	Dept. of Energy
ESF13	Public safety and security	Dept. of Homeland Security/Justice
ESF14	Cross-sector business and infrastructure	Dept. of Homeland Security/Cybersecurity and Infrastructure Security Agency (CISA)/ FEMA
ESF15	External affairs	Federal Emergency Management Agency

[1] *Source:* FEMA[38]

in vector control after disasters either directly or indirectly. Sometimes FEMA, in coordination with the CDC, may directly contract with vendors for aerial or ground spraying of pesticides (all paid up front), but most of the time this work is done by reimbursement. All mosquito control activity after disasters is potentially eligible as public assistance,

Category B Emergency Protective Measures, as outlined in the FEMA Public Assistance Program and Policy Guide (PAPPG), FP 104–009–2.[36,37] Since FEMA does not pay for everything related to vector control after a disaster, it is important for public health and local elected officials to review and understand the types of expenses that are reimbursable and the documentation required. For example, FEMA does not automatically pay all activities, such as overtime, after an event. Further, since mosquito season and hurricane season overlap, much of mosquito control activities, equipment, and chemicals during this time period are considered "normal" expenses unrelated to the disaster. Therefore, FEMA only provides public assistance for the *increased* cost of mosquito abatement. In other words, the amount that exceeds the average amount based on the last 3 years of expenses for the same time period. In order to receive federal funding, good record keeping is essential. Daily truck mileage logs, spray equipment logs, overtime and other records must be maintained annually to demonstrate an increase in workload and materials due to an incident.[37] If the reimbursement request is based on potential for disease transmission, a mosquito control program should be prepared to show that there are higher levels of disease-transmitting mosquitoes in the disaster area following the event as compared to pre-disaster surveillance data. For example, after Hurricane Katrina, mosquito collections along the Mississippi Gulf Coast revealed that *Culex nigripalpus* was the predominant species collected in the affected area (71.2% of all mosquitoes trapped). Since *C. nigripalpus* is a known vector for WNV, the case for increased disease threat in the area was easily made (for more information, see section, *Mississippi Department of Health Katrina Experience* in Chapter Six). An outline for the process of requesting disaster mosquito control justification, surveillance, and required data are found in the PAPPG Appendix G, *Mosquito Abatement*.[36]

Military Aerial Spraying Capability

Entomology, and specifically vector control, is very important in the success of military campaigns. In fact, vector-borne diseases have often severely reduced the fighting capability of armed forces, even suspending or canceling wars.[39] Military entomologists are acutely aware of the disease threat to troops from arthropod pests, and have remained an integral part of the Department of Defense (DoD) since they were deployed with excellent results to combat malaria in the South Pacific during World War II.[40,41] Military entomologists provide equipment and subject matter expertise for disease surveillance and vector control during troop deployments, but may also assist in reestablishment of normal living conditions after wars or natural disasters.[42] For example, entomologists helped maintain troop health during World War II and in subsequent conflicts by application of insecticides from aircraft to areas with high mosquito activity. Aerial applications of pesticides can rapidly reduce numbers of potential insect vectors across large areas in a relatively short period of time. There are reports of large decreases in mosquito populations following spraying and decreases in virus infection rates in vector mosquitoes. Unlike ground spraying with truck-mounted spray units or backpack sprayers, aircraft can access developed and undeveloped areas that are prone to arthropod-borne disease outbreaks. As an example of aerial spraying success, mosquito landing counts were reduced by ≥ 93% for 4 days at Ft. Stewart, Georgia, in 1960 following aerial spraying.[43]

THREE PHASES FOR EMERGENCY MOSQUITO SPRAYING[44]

1. **Premission:** finding a place to work, relocating people and equipment, mapping, preflight surveillance, product (pesticide) acquisition, and public relations
2. **Mission:** mosquito surveillance and identification, adult mosquito control with aerial ultra-low volume (ULV) application equipment
3. **Postmission:** evaluating spray efficacy, final mapping, and reports to customer

The U.S. Air Force is responsible for large-area aerial pesticide application capability to control disease vectors, pest organisms, and undesirable/invasive vegetation on DoD installations, or over nonmilitary lands during declared emergencies. In several publications, Maj. Mark Breidenbaugh and colleagues have reviewed the USAF role in aerial application of pesticides,[40,45,46] and much of the following discussion comes from those papers. The Air Force Aerial Spray Unit (AFASU) traces its history back to applications of DDT by plane during the later stages of World War II. After the end of the war, a unit called the Special DDT Flight was created, but was soon transformed to the Special Aerial Spray Flight (SASF) in 1947 when the Air Force became a separate armed service. After more than 25 years at Langley Air Force Base, Virginia, the SASF was transferred from the active Air Force to the Air Force Reserve in 1973, but prior to this transfer, the SASF had sprayed for mosquitoes, Japanese beetles, and fire ants in various locations at the request of the Army, Navy, and other federal agencies. Relocated to Rickenbacker Air Force Base, Ohio, the unit was soon renamed the Spray Branch of the 907th Tactical Airlift Wing. In 1986, the Spray Branch began to transition from C-123 airplanes (Figure 6.12) to C-130A airplanes and developed the modular aerial spray system for use

Figure 6.12 C-123 aircraft previously used for aerial spraying.

Source: From the United States Air Force

Figure 6.13 C-130 aircraft adapted for aerial spraying.

Source: From the United States Air Force

in C-130E and H airplanes (Figure 6.13). In 1991, the aerial spray mission was assigned to the 901th Airlift Wing at the Youngstown Air Reserve Station, Ohio, and currently trains for a primary wartime mission of protecting deployed troops from arthropod-borne illness. However, this unit has often been utilized in nonmilitary emergency aerial spray responses since that time (Table 6.3). One good thing about the Air Force Aerial Spray Unit being placed at a reserve station (instead of an active duty location),

Table 6.3 Aerial Spraying Projects Conducted by the Military.

Nonmilitary emergency deployments by the AFASU
after transition to the U.S. Air Force Reserve[40]

Year	Location	Health threat	Coverage (acres)
1973	Panama	Equine encephalitis	37,600
1975	Guam	Dengue fever	157,530
1978	Azores	Japanese beetles	8,700
1983	Minnesota	Equine encephalitis	525,000
1985	Idaho	Grasshoppers	718,100
1987	Puerto Rico	Dengue fever	177,000
1989	South Carolina	Hurricane Hugo mosquito control	855,500
1992	Florida	Hurricane Andrew mosquito control	279,170
1999	North Carolina, Virginia	Hurricane Floyd mosquito control	1,700,000
2005	Louisiana, Texas	Hurricanes Katrina and Rita mosquito control	2,880,622

is that the members are not moved every few years, allowing aircrews to develop a strong skill set necessary to conduct aerial spraying.[45]

How the Military Spray System Works

To facilitate ease of handling, use, and repair, a modular aerial spray system (MASS) was developed for use with the C-130H airplane, having a maximum 2,000-gallon capacity for liquid materials, which can be rolled on or off the airplane in approximately 30 min. Functional in a variety of configurations, the MASS is useful for such applications as ultra-low volume (ULV) adult mosquito sprays (adulticiding), mosquito liquid larvicide sprays, herbicide applications, and oil dispersants for emergency cleanup of oil spills. Ultra-low-volume sprays create an aerosol cloud of small discrete droplets that drift through the air and flying insects are killed by contacting these droplets. These sprays do no good once they hit the ground. For this reason, the flight period of the target pests is one of the most important planning factors for ULV spray missions. The current configuration for AFASU mosquito adulticiding uses the MASS with booms placed through the fuselage doors. Those booms are fitted with flat fan nozzles positioned perpendicular to the slipstream of the aircraft for maximum shear and atomization of pesticide. This is especially important since the diameter of a droplet that effectively adheres to a mosquito is 10 to 25 μm. Droplets too big or too small will not effectively contact flying pests such as mosquitoes.

Air Force Response to Hurricane Katrina

The AFASU played an integral role in vector control in Louisiana after Hurricane Katrina.[47] The unit was officially assigned to Joint Task Force Katrina on September 9, 2005, wherein two spray aircraft, a spare aircraft, and three crews set up at Duke Field, Florida, near Eglin AFB on the Florida panhandle. Duke Field was chosen because it was the closest fully functioning military base with the logistical capability of supporting the C-130H spray aircraft. A total of 53 personnel were involved with the flying, entomology, maintenance, administration, communication, and life support issues relating to the AFASU response. In September, the full scope of damage and human suffering from Katrina was still unknown. News agencies reported potential human fatalities as high as 10,000 inside New Orleans, incidents of gunfire aimed at rescue aircraft, and EPA water tests showing *E. coli* levels 10 times above safe levels. Rapid mosquito and fly development was anticipated as temperatures reached 90°F during the day, and only cooled into the upper 70s (°F) at night. Filth flies were considered to be the insect vector of immediate health importance since they could develop quickly in flooded New Orleans (e.g., fecal contamination, muck, trash, animal carcasses)[48] and move between filth and human food, potentially transferring harmful bacteria mechanically in the process. There was also some concern that West Nile virus (WNV) was circulating in mosquito and bird populations. In the week ending August 30, 2005, 40 cases of WNV had been reported in Louisiana, including 4 deaths. This was already more cases in Louisiana than in all of 2004. The hurricane had created enough damage to expose the population to nuisance mosquito biting and possible disease transmission. In particular, the absence of electricity, physical damage to structures, and the use of temporary shelters by an estimated 200,000 displaced people combined to make these individuals vulnerable to mosquito attack. Another aspect of mosquito biting was the threat to responders—rescue, cleanup, and utility repair crews, as well as law enforcement personnel. Therefore, mosquito and fly control became the highest priorities in public health protection after the hurricane.

Conducting aerial spray operations in and around New Orleans was complicated. The AFASU had previously conducted spray operations during federally declared disasters, but the situation in New Orleans was particularly challenging. Entomologists from the Louisiana Department of Health, CDC, and Air Force conducted surveillance in and around New Orleans, performing mosquito landing counts, surveys for filth flies, and direct consultation with bivouacking Army and Navy preventative medicine troops.

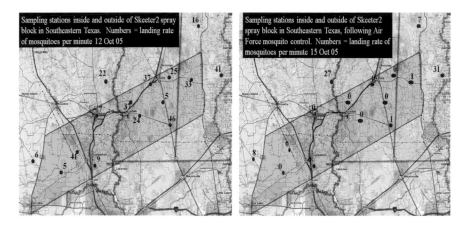

Figure 6.14 Mosquito landing rates in spray zone before and after spraying.

Source: From the United States Air Force

Since telephone and e-mail services were sporadic (at best) and troops continued to be redeployed to new areas, this surveillance was the primary and most effective means of evaluating insect vector potential in the area. Aerial spraying was the only alternative for mosquito control since driving in New Orleans was difficult, if not impossible, and since many roads were flooded or obstructed by debris. Further, traffic flow was severely impeded by military and police checkpoints located almost everywhere.

During the Hurricane Katrina spraying mission, a total of 1,942,607 acres in 12 Louisiana parishes was sprayed by the USAF. There was considerable evidence of good control (Figure 6.14), even from non-Air Force sources. One environmental toxicology group taking samples from New Orleans from September 16 to 18 reported "minimal adult mosquito activity in the area."[48] In addition, in Acadia Parish, mosquito control personnel measured mosquito landing rates on September 29 and found an average of 49 per minute. Much of the parish was sprayed that night and there was a 88% reduction in landing rates the next day in sprayed areas, while in unsprayed areas average landing rates increased to 91 per minute. The area was treated again on September 30, this time yielding 99% reduction in landing rates the next day.[47]

Mississippi Department of Health Katrina Experience

Hurricane Katrina devastated the 90-mile Mississippi Gulf Coast on August 29, 2005 (Figure 6.15). The event was a total ecological disaster due to salt water (storm surge) washing ashore for several miles inland. At one spot near Waveland, Mississippi, this author observed the debris

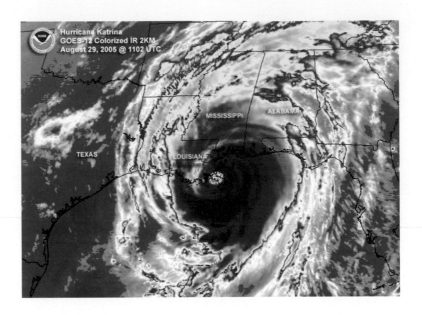

Figure 6.15 Hurricane Katrina devastated the Mississippi Gulf Coast.

Source: From the National Oceanic and Atmospheric Administration

line on the south side of Interstate 10, representing the water's height, at approximately 25 ft high. This was 7 miles inland! Further, this meant that everything in the storm surge's path was completely inundated with salt water, a lot of which was washed back out to sea. Accordingly, no insect or rodent problems occurred during the first week or so after the storm; however, new mosquito, filth fly, and rodent breeding began shortly after that (Table 6.1).

Four days after the storm, I went to the disaster zone for a meeting with Mississippi Emergency Management personnel and the Mississippi Department of Health (MDH) incident commander to plan the vector control response (the Incident Command System was utilized after Katrina—see text box and Figure 6.16). Interestingly, at that time, no mosquitoes were seen, nor even many birds, as these were apparently all "blown out" of the area by the storm. By day 11, a team of entomologists (two from the MDH and seven from the CDC) was in place at a local high school to coordinate vector surveillance and control (Figure 6.17). We ran 15–18 CO_2-baited CDC mosquito traps each night in the six lower Mississippi counties (three counties along the coast, three above them) and ran the

During emergencies all states use the Incident Command System

INCIDENT COMMAND SYSTEM: COMMAND STAFF & GENERAL STAFF

Figure 6.16 The Incident Command System is used in disaster responses nationwide.

Figure 6.17 Makeshift mosquito identification lab set up in a local high school after Hurricane Katrina.

Table 6.4 Mississippi Department of Health Battle Plan for Vector Control
after Hurricane Katrina, Fall 2005.

Disaster vector control objectives

1. Establish systematic mosquito surveillance in the affected counties.
2. Test pools of mosquitoes for West Nile virus and other arboviruses.
3. Intervene where necessary to control mosquito hotspots.
4. Provide mosquito repellent free of charge to residents and responders.
5. Send out teams of MDH environmental health personnel to identify and larvicide mosquito breeding sites.
6. Work with local mosquito control entities to help get them back online and functioning.

traps daily for several weeks after the storm. Our objectives for post-storm vector control were fairly simple (Table 6.4), and included helping local mosquito control agencies reestablish services, and performing mosquito surveillance, larviciding, and adulticiding where necessary. County environmental health specialists (EHSs) were enlisted to aid MDH entomologists as foot patrols for larviciding. In addition, MDH personnel helped write action request forms requesting Federal Emergency Management Agency (FEMA) assistance for increased ground spraying in the six lower counties.

Destruction from the storm was almost unbelievable. Thousands of homes were totally destroyed, leaving nothing but foundation or slabs. Others, a few miles inland, were partially destroyed with roofs, windows, and doors gone. Virtually all infrastructure in the coastal counties collapsed; electric lines were all downed; and sewer and water pipes broke, sending raw sewage over the ground surface. Roads, where passable, were littered with debris, causing extremely slow traffic and flat tires. People slept either outside or inside their partially destroyed homes. Heat and humidity were unbearable. For the first week or two, people lived on military meals ready-to-eat or other foods which they stored on ice in coolers (FEMA distributed bags of ice) (Figure 6.18). Sanitation lagged, with people using makeshift latrines or simply piles of rubbish for bathrooms. Filth flies contaminated food and food surface areas like spoons, forks, and knives (Figure 6.18). Responders lived in tent cities, but many tents were open, allowing unrestricted filth fly access (Figure 6.19).

One of the most significant pest control issues after the hurricane was the debris piles. The enormity of such piles was staggering, as they contained thousands of people's houses and property that had been destroyed and scattered about by the storm. Some piles reached to the

Figure 6.18 After Hurricane Katrina many people ate meals out of coolers under primitive and unsanitary conditions. Note flies on the knife and cigarette.

Source: Photo courtesy Dr. Wendy C. Varnado).

Figure 6.19 Responders stayed in tent cities after Hurricane Katrina.

Source: Photo courtesy Dr. Wendy C. Varnado

Figure 6.20 Huge debris pile along Mississippi Coast after Hurricane Katrina. Note the dump truck in the foreground.

Source: Photo copyright 2005 by Jerome Goddard, Ph.D.

top of telephone poles in height; others looked like small mountains (Figure 6.20). FEMA paid contractors with heavy equipment to clear roads and push debris from the storm into piles along the side of the road for pickup. These piles contained myriad food and breeding sites for mosquitoes, flies, rodents, and other vermin, including water in artificial containers, household food from freezers and refrigerators, and dead wild and domestic animals. One veterinary clinic I saw had piles of pet food outside on the street—perfect feeding stations for rodents. Removal of debris piles took several months to over a year in some areas, leading to development of other, more semipermanent pest problems, like brown widow spiders (Figure 6.21).

The need for rodent control after the hurricane in Mississippi was minimal (in contrast to New Orleans where rodent problems were significant). The health department coordinated with several private pest control companies to place rodenticides in secured bait stations throughout affected areas. These same private companies were responsible for cleanup and removal of unused baits.

Figure 6.21 Brown widow spiders moved into debris piles after several months. Egg sacs shown here.

Figure 6.22 Wendy Varnado, Mississippi Department of Health, conducts larval sampling at home destroyed by Hurricane Katrina.

Our first line of mosquito control was larviciding, based upon surveillance data (Figure 6.22); however, at times we larvicided even without any surveillance information. EHS larviciding teams were assembled, armed with *Bacillus thuringiensis israeliensis* (BTI) briquettes (some with Altosid®), and told to toss them into any sources of standing water that might serve

Figure 6.23 Larviciding water source with BTI dunk.

Source: Photo courtesy Rosella M. Goddard

Figure 6.24 Warning sign to looters.

Source: Photo courtesy Dr. Wendy C. Varnado, Mississippi Department of Health

as breeding grounds (Figure 6.23). Copies of labels and Material Safety Data Sheets (MSDSs) were available to give residents if they had questions. Treatment of standing water included both public and private properties. Due to public fears of looters, who were threatened with shooting (Figure 6.24), EHS teams were appropriately identified by MDH shirts

Figure 6.25 Example of clearly marked responder shirts.

Source: Photo courtesy Dr. Wendy C. Varnado, Mississippi Department of Health

and badges (Figure 6.25). They were essentially given free rein to treat for mosquitoes anywhere they saw a breeding site. Note: Extreme situations sometimes require extreme measures. Treating private property for mosquitoes is not ordinarily permitted by state and local laws without an imminent health hazard present. This was a moot point anyway, because after Katrina, distinguishing public lands such as roads from homesites, or anything else for that matter, was extremely difficult. The landscape was nothing but a sea of junk and debris. This was precisely why BTI products (ordinarily over-the-counter, general-use pesticides) were chosen for the larviciding effort—it would be hard to misuse such products and cause any real harm to people or the environment. Areas larvicided were marked on a map each evening to prevent overtreating and to help plan the next day's work schedule (Figure 6.26). During the larviciding effort, over 20,000 mosquito dunks and 4,800 Altosid briquets were applied to a 62-square-mile area in the three coastal counties.[49] Applications were made primarily to damaged swimming pools, garbage and debris piles holding water, and roadside ditches. In addition to the MDH larviciding efforts, we became aware of other military or FEMA groups also involved in mosquito larviciding. For example, an Emergency Support Function No. 8 team from Florida reported to our incident commander that they had applied more than 1,000 methoprene briquets along the Mississippi Gulf Coast.[49]

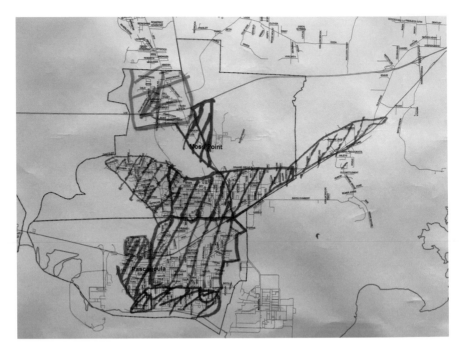

Figure 6.26 Larviciding map used after Hurricane Katrina.

Source: Photo courtesy Jerome Goddard, Ph.D.

Approximately 2 weeks after hurricane landfall, due to increasing mosquito numbers in CDC light traps, MDH entomologists requested help from FEMA for an aerial spraying of insecticide to reduce nuisance mosquito biting on residents (living in partially or totally destroyed dwellings) and responders. FEMA subsequently issued a contract with a private contractor for one application of Dibrom® insecticide. The organophosphate naled (brand names both Dibrom and Trumpet) is the active ingredient most often used in post-hurricane or post-flooding aerial applications of adulticides due to its favorable characteristics of density, efficacy, and the short-lived nature of the product in the environment.[50] Results were dramatic, with over 91% control in entire counties in one night (Figures 6.27 and 6.28). No complaints or medical or environmental problems from the increased ground spraying and aerial spraying were reported.

During the MDH response to Hurricane Katrina, a total of 60,533 mosquitoes was collected, representing 34 species in 9 genera. The predominant species collected was *Culex nigripalpus*, comprising 71.2% (43,072/60,533) of all mosquitoes trapped, and was widespread in all counties. Other commonly collected species were *Ae. atlanticus/tormentor*

Figure 6.27 Aerial spraying was conducted in Mississippi after Hurricane Katrina.
Source: Photo courtesy Dr. Wendy C. Varnado, Mississippi Department of Health

**Aerial Spray Results – 1 Day after Spraying
Ten Traps Combined**

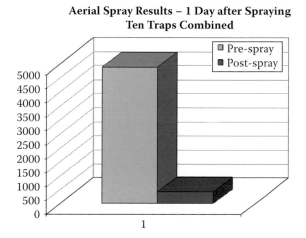

Figure 6.28 Aerial spraying after Hurricane Katrina was very successful in reducing mosquito numbers.

and *Ps. ferox*. At the time, high numbers of *Cx. nigripalpus* were worrisome in light of its vector competence for WNV. However, subsequent testing by the CDC of almost 50,000 mosquitoes from this sample yielded no positives for either WNV or EEE virus.

USING THE INCIDENT COMMAND SYSTEM FOR DISASTERS

The Incident Command System (ICS) was first developed to provide federal, state, and local governments, as well as private and charitable entities, with a consistent framework for the preparation for, response to, and recovery from any incident or event, regardless of the size, nature, duration, location, scope, or complexity. ICS utilizes core concepts of interoperability, shared terminology, and technologies, and especially unified command to manage incidents such as natural disasters. The ICS concept has evolved and been adapted into a national system. In 2003, President Bush directed the Secretary of Homeland Security to develop the National Incident Management System (NIMS). NIMS provides a systematic, proactive approach to guide departments and agencies at all levels of government, nongovernmental organizations, and the private sector to work seamlessly to prevent, protect against, respond to, recover from, and mitigate the effects of incidents, regardless of cause, size, location, or complexity, in order to reduce the loss of life and property and harm to the environment.

The Katrina Experience

After a monumental disaster like Hurricane Katrina, responders of all types flood into the area—governmental, private, faith-based, and others. Almost immediately, it becomes evident that no one seems to be in charge of the response, with people coming and going and performing rescue or cleanup tasks that *they believe* to be most important. In many areas, chaos ensues because each responder group has its own funding source, chain of command, and communication network. NIMS was designed to eliminate this problem by unifying the command structure after an incident and coordinating the response. However, most people agree that NIMS did not perform well during the Hurricane Katrina response, especially in New Orleans, and exhibited serious flaws. Several governmental and nongovernmental entities issued reports on the response to Hurricane Katrina. Three particularly telling ones were: *A Failure of Initiative* from the U.S. House of Representatives, *Hurricane Katrina: A Nation Still Unprepared* issued by the U.S. Senate, and *The Federal Response to Hurricane Katrina: Lessons Learned from the White House.* Each took the position that change in how disaster response organizations function is necessary. Words like *initiative* and *transformation* appeared throughout the texts, and although they differed in prescriptions, the basic point of each report was the need to transform the current system. NIMS seemed to work better in Mississippi after Katrina, but in preparation for future incidents, all Mississippi Department of Health personnel were required to take NIMS training online after 2005. All agencies in all states should do the same to facilitate disaster management. The more people who receive NIMS training, the more coordinated and efficient the disaster response will be.

Overall, government and private industry response to Hurricane Katrina was deemed a great success. Nuisance mosquito biting and filth fly pests were abated using a combination of approaches, and valuable lessons were learned about multiple agency coordination and judicious use of pesticides to protect public health.

References

1. CDC. A national framework for the prevention and control of vector-borne diseases in humans. In: U.S. Centers for Disease Control; 2020:16 pp. www.cdc.gov/ncezid/dvbd/framework.html.
2. WHO. World malaria report, 2020. In: World Health Organization, Geneva, Switzerland; 2020. www.who.int/teams/global-malaria-programme/reports/world-malaria-report-2020.
3. Gratz NG. Emerging and resurging vector-borne diseases. *Ann Rev Entomol.* 1999;44:51–75.
4. Molyneux DH. Patterns of change in vector-borne diseases. *Ann Trop Med Parasitol.* 1997;91:827–839.
5. Anonymous. Dengue more prevalent than thought. *Science (News Focus).* 2013;340:127.
6. Bhatt S, Gething PW, Brady OJ, et al. The global distribution and burden of dengue. *Nature.* 2013;496(7446):504–507.
7. CDC. Locally acquired dengue—Key West, Florida, 2009–2010. In: CDC, MMWR. 2010;59:577–581.
8. Leal A, Raoke E. The 2009 dengue outbreak in the Florida Keys. In: American Mosquito Control Association, Wing Beats Magazine. 2010;Spring Issue:7–10.
9. Dennis DT. Plague as an emerging disease. In: Scheld WM, Craig WA, Hughes JM, eds. *Emerging Infections.* Vol. 1. Washington, DC: ASM Press; 1998:169–183.
10. CDC. Human plague—United States, 1993–1994. In: MMWR. 1994;43:242–243.

11. CDC. Summary of notifiable infectious diseases and conditions—United States, 2015. In: CDC, MMWR. 2017;64(53):1–144.
12. CDC. Summary of notifiable diseases. In: National Notifiable Diseases Surveillance System (NNDSS); 2018. wwwn.cdc.gov/nnds/infections-tables-html.
13. Enserink M. Has leishmaniasis become endemic in the U.S.? *Science (News Focus).* 2000;290:1881–1883.
14. Kuehn BM. CDC estimates 300,000 U.S. cases of Lyme disease annually. *JAMA.* 2013;310:1110.
15. Ginsberg HS, Faulde MK. Ticks. In: Bonnefoy X, Kampen H, Sweeney K, eds. *Public Health Significance of Urban Pests.* Copenhagen: WHO Regional Office for Europe; 2008:303–345.
16. Gorenflot A, Moubri K, Precigout E, Carcy B, Schetters TP. Human babesiosis. *Ann Trop Med Parasitol.* 1998;92:489–501.
17. Walker DH. Emerging human ehrlichioses—recently recognized, widely distributed, life-threatening, tick-borne diseases. In: Scheld WM, Armstrong D, Hughes JM, eds. *Emerging Infections.* Vol. 1. Washington, DC: ASM Press; 1998:81–91.
18. Thomford JW, Conrad PA, Telford SR, III, et al. Cultivation and phylogenetic characterization of a newly recognized human pathogenic protozoan. *J Infect Dis.* 1994;169:1050–1056.
19. Dumler JS, Bakken JS. Human ehrlichioses: newly recognized infections transmitted by ticks. *Ann Rev Med.* 1998;49:201–213.
20. Pritt BS, Sloan LM, Johnson DKH, et al. Emergence of a new pathogenic Ehrlichia species, Wisconsin and Minnesota, 2009. *New Engl J Med.* 2011;365(5):422–429.
21. Medlock JM, Leach SA. Effect of climate change on vector-borne disease risk in the U.K. *Lancet Infect Dis.* 2015;15:721–730.
22. IPCC. Climate change 2014: impacts, adaptation, and vulnerability. In: Intergovernmental panel on climate change, working group II contribution—changes to the underlying scientific/technical assessment; 2014. www.ipcc.ch/report/ar5/wg2/.
23. Lafferty KD. The ecology of climate change and infectious diseases. *Ecology.* 2009;90:888–900.
24. Shuman EK. Global climate change and infectious diseases. *N Engl J Med.* 2010;362:1061–1063.
25. Scheld WM, Armstrong D, Hughes JM. *Emerging Infections.* Vol. 1. Washington, DC: ASM Press; 1998.
26. Chomel BB, Belotto A, Meslin FX. Wildlife, exotic pets, and emerging zoonoses. *Emerg Infect Dis.* 2007;13:6–11.
27. CDC. CDC health information for international travel; 2010. wwwnc.cdc.gov/travel/content/yellowbook/home-2010.aspx.
28. CDC. Traveler's health page; 2010. wwwnc.cdc.gov/travel/.
29. James B. Disaster preparedness and mitigation. In: UNESCO Publication Number SC/BES/NDR/2007/H/1; 2007:48 pp.
30. Scott HG. Emergency vector-borne disease control: an orientation for environmental health personnel. *J Environ Health.* 1964;26:21–28.
31. Kampen H, Schaffner F. Mosquitoes. In: Bonnefoy X, Kampen H, Sweeney K, eds. *Public Health Significance of Urban Pests.* Copenhagen: World Health Organization; 2008:347–386.

32. Mount GA, Biery TL, Haile DG. A review of ultralow-volume aerial sprays of insecticide for mosquito control. *J Am Mosq Control Assoc.* 1996;12:601–618.

33. Carney RM, Husted S, Jean C, Glaser C, Kramer V. Efficacy of aerial spraying of mosquito adulticide in reducing incidence of West Nile Virus, California, 2005. *Emerg Infect Dis.* 2008;14(5):747–754.

34. Elnaiem DE, Kelley K, Wright S, et al. Impact of aerial spraying of pyrethrin insecticide on *Culex pipiens* and *Culex tarsalis* (Diptera: Culicidae) abundance and West Nile virus infection rates in an urban/suburban area of Sacramento County, California. *J Med Entomol.* 2008;45(4):751–757.

35. Eldridge BF. Strategies for surveillance, prevention, and control of arboviral diseases in Western North America. *Am J Trop Med Hyg.* 1987;37 Suppl.:77S–86S.

36. FEMA. Public assistance program and policy guide. In: U.S. Department of Homeland Security, FP 104-009-2; 2018. www.fema.gov/media-library/assets/documents/111781.

37. McAllister, J.C., Madson SL. Federal assistance for mosquito abatement post-disaster or during disease outbreaks. *J Am Mosq Control Assoc.* 2020;36(2S):74–77.

38. FEMA. National response framework. 4th ed. Washington, DC: U.S. Department of Homeland Security; 2019. www.fema.gov/media-library/assets/documents/11791.

39. Pages F, Faulde M, Orlandi-Pradines E, Parola P. The past and present threat of vector-borne diseases in deployed troops. *Clin Microbiol Infect.* 2010;16:209–224.

40. Breidenbaugh M, Haagsma K. The U.S. Air Force Aerial Spray Unit: A history of large area disease vector control operations, WWII through Katrina. *Army Med Dept J.* 2008;April–June Issue:1–8.

41. Cushing E. *History of Entomology in World War II.* Washington, DC: Smithsonian Institution; 1957.

42. Cope SE, Schoeler GB, Beavers GM. Medical entomology in the United States Department of Defense: challenging and rewarding. *Outlooks Pest Manag.* 2011;June issue:129–133.

43. Dowell FH. An examination of the United States Air Force aerial spray operations. *Mosq News.* 1965;25:209–216.

44. Boze BGV, Markowski DM, Bennett D, Williams MG. Preparations and activities necessary for aerial mosquito control after hurricanes. *J Am Mosq Control Assoc.* 2020;36 (2S):90–97.

45. Lindroth EJ, Breidenbaugh MS, Stancil JD. U.S. Department of Defense support of civilian vector control operations following natural disasters. *J Am Mosq Control Assoc.* 2020;36 (2S):82–89.

46. Breidenbaugh M, Haagsma K, Walker W, Sanders D. Post-Hurricane Rita mosquito surveillance and the efficacy of Air Force aerial applications for mosquito control in east Texas. *J Am Mosq Control Assoc.* 2008;24:327–330.

47. Breidenbaugh M, Haagsma K, Olson GS, et al. Air Force aerial spray operations to control adult mosquitoes following hurricanes Katrina and Rita. In: American Mosquito Control Association, Wing Beats Magazine. 2006;Summer issue:7–15.

48. Presley SM, Rainwater TR, G.P. A, et al. Assessment of pathogens and toxicants in New Orleans, LA following Hurricane Katrina. *Environ Sci Technol.* 2006;40:468–474.

49. Goddard J, Varnado WC. Disaster vector control in Mississippi after Hurricane Katrina: lessons learned. *J Am Mosq Control Assoc.* 2020;36:56–60.
50. Trumbetta JM, Placentia V, Connellly PH. Meeting increased demand for mosquito adulticides containing the active ingredient naled following hurricanes and tropical storms. *J Am Mosq Control Assoc.* 2020;36 (2S):98–102.

chapter seven

Operational Research Opportunities in Public Health Entomology

Background and Purpose of Research in Public Health Entomology

Federal and state-level policymakers and public health practitioners must implement public health approaches as well as medical treatments for the overall good of society. However, research into these public health approaches lags far behind research on new therapies, biotechnology, and medical devices. In fact, a study in 2010 showed that biomedical research gets about 90% of governmental funding, compared to about 10% for public health research (this likely changed after the covid pandemic).[1] Scientific research is important in public health, mainly for providing useful information for community health programs and results of interventions. Such research sheds light on diseases and conditions in the state or territory and associated demographics. Health officers would be ill-advised to initiate expensive programs for diseases or other problems in their state without knowing what those diseases are, where they are, and in what numbers. This is *descriptive research*. In addition, basic research is sometimes needed to identify new or emerging pathogens in one's state, their ecology, and modes of transmission. For example, the author helped with a survey of ticks found on jackrabbits in Utah, looking for a new disease agent (Figure 7.1). In some cases, new or poorly known disease studies require *experimental research*—hypothesis testing—but the purpose is the same: shedding light on the public health threats and conditions within the state or territory.

Types of Research

Historically, entomological research projects conducted by state health departments included practical aspects such as mosquito breeding potential in irrigated farmlands, insecticide resistance in mosquito vectors, and ecology and control of filth flies, ticks, biting midges, and stable flies.[2,3] Modern public health research can consist of surveys (including

Figure 7.1 Survey for new disease agent in ticks found on jackrabbits in Utah.

Source: Photo copyright 2013 by Jerome Goddard, Ph.D.

surveys of vectors for insecticide resistance),[4] vector surveillance, or other descriptive projects, as well as experimental research concerning the ecology or biology of various disease agents and their modes of transmission (Figures 7.2–7.4). However, for the most part, research at public health departments is not basic academic research, for example, learning previously unknown facts just for the sake of learning those facts. For the entomology program, basic research consists of tick, mosquito, flea, or other vector surveys to ascertain species diversity, location, and seasonality within the state, or ecological studies of vector-borne diseases to determine dynamics involved in transmission of those diseases. Practical research might also be conducted on insecticide efficacy to survey which pesticides work best to control infestations or abate nuisances (Figures 7.5–7.7 depict results from insecticide tests for ticks and bed bugs).[5] For examples of research at the Mississippi Department of Health, we tested efficacy of repellents (Figure 7.8), surveyed the state for ticks,[6–9] mosquitoes,[10–14] and

Figure 7.2 Surveys for ticks involve dragging a white cloth through the woods.

Figure 7.3 Survey for mosquito larvae in cemetery vases.

Source: Photo Dr. Gail Moraru and the Mississippi State University Extension Service

Figure 7.4 Survey for mosquito larvae in the purple pitcher plant.

Source: Photo copyright 2005 by Jerome Goddard, Ph.D.

Figure 7.5 Efficacy testing of insecticide products for ticks.

Figure 7.6 Evaluation of insecticides against bed bugs.

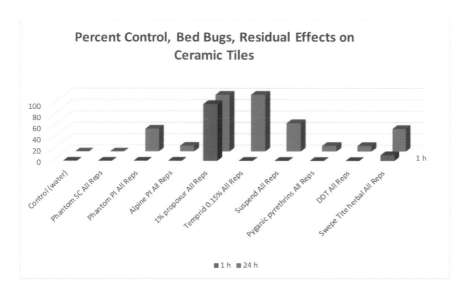

Figure 7.7 Example of an efficacy test of insecticides against bed bugs.

Note: the proper context surrounding these data are not provided such as strain of bed bug population tested, dilutions, application rates, etc.)

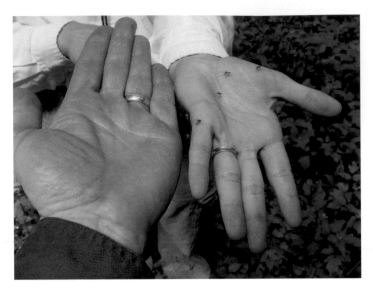

Figure 7.8 Field testing of mosquito repellents. Hand on the left was treated with DEET-based repellent.

Source: Photo copyright 2004 by Jerome Goddard, Ph.D.

sand flies.[15] In collaboration with scientists at Mississippi State University, we surveyed ticks[16] and cattle[17] for the agent of Lyme disease[18] and spotted fever group (SFG) rickettsiae, and especially the newly recognized pathogen, *Rickettsia parkeri*.[19,20] In addition to these descriptive research projects, we conducted experimental research on tick ecology and relative risk of acquiring ticks in the woods.[21-23]

DON'T FORGET THE INSTITUTIONAL REVIEW BOARD

For most biomedical and behavioral research involving humans, an institutional review board (IRB) or ethical review board must approve, monitor, and review research projects. IRBs are mandated in the United States by Title 45 CFR Part 46 and regulated by the Office for Human Research Protections (OHRP) within the Department of Health and Human Services. IRB approval for a research project is necessary to protect the rights and welfare of the research subjects. Health departments may or may not have an official IRB, and often contract this out or otherwise use an outside review board. Nonetheless, IRB approval is still needed for any research involving human subjects, even at a nonacademic institution like a health department (this may include even simple research projects like questionnaires). Entomologists sometimes forget the need for IRB approval and conduct research projects that actually should require IRB oversight. For example, medical entomologists may wish to feed mosquitoes, ticks, or bed bugs on themselves as part of colony maintenance or an experiment; however, IRB approval may still be needed even if the researcher experiments on himself or herself.

As for practical or operational aspects of public health research, consider the following example. In 2005, research by personnel at the Sacramento and Yolo Mosquito Abatement District demonstrated that aerial spraying with pyrethrin insecticide reduced the transmission intensity of WNV and decreased the risk of human infection.[24] Another paper related to that same research project demonstrated that the odds of infection with WNV were six times higher in untreated areas than in treated areas.[25] These studies provided much needed objective information about the usefulness of mosquito control in preventing/controlling WNV infections; that is, they demonstrated *effectiveness* of a public health intervention.

Funding Sources and Examples of State Labs

Some health departments are well funded with state resources for research and have state-of-the-art laboratories. For example, the entomology program at the Mississippi Department of Health is located within an almost new public health laboratory with state-of-the-art laboratory facilities available, including access to a BSL-3 lab. They have equipment available for both conventional and real-time PCR analysis of mosquito samples. In Tennessee, a dedicated vector-borne disease laboratory (VDL) is available for research projects within the Vector-Borne Disease Branch of the Communicable and Environmental Diseases Section at the Tennessee Department of Health (TDH), and serves as a great model for other states.[1] Their laboratory space consists of a large main room (approximately 800 ft^2) where general immunoassay and molecular assay work is conducted. VDL recently acquired new space for cell culture work and a "fellow" room for research fellows, interns, and graduate students who work on various projects. An insectary is also available for bringing arthropods back from the field to distinguish by species. Equipment available at the TDH VDL includes two Revco Ultima II Upright—80°C freezers and one—20°C freezer, two Forma Model 3960 Reach-in Incubators for growth and maintenance of mosquito colonies, a Spectronic Biomate for spectrophotometry, a Genvac miVac Concentrator for preparing samples for sequencing, a New Brunswick Classic Model C25 Floor Model Incubator Shaker for bacterial growth during cloning if direct sequencing proves problematic, a heating circulator water bath, a variable intensity transluminator, a photodocumentation system, a fluorescent microscope, an enzyme-linked immunosorbent assay (ELISA) plate reader, various equipment and supplies for electrophoresis, and a conventional thermal cycler. The combined lab facilities sections provide a Qiagen BioRobot 9604 for extraction of DNA from large numbers of mosquito samples, real-time and conventional thermal cyclers for PCR, and an automated sequencer to sequence samples in-house.

Biomedical research funding for public health entities is often derived differently than that of university academic research (which is from agencies like the National Institutes of Health or National Science Foundation). Grants to public health departments are frequently noncompetitive and originate from the Centers for Disease Control, federal preventive medicine block grants, other public money, or perhaps private foundations. In some cases, state legislatures earmark or line-item portions of the health department budget for research on particular diseases or conditions. As examples of noncompetitive grants, when I was affiliated with the Mississippi Department of Health, we received federal grants totaling several million dollars designated for mosquito control and West Nile virus surveillance and research. While a grant proposal was required for each one, the money was already designated for our state. Although we used most of this money for establishment of arboviral surveillance infrastructure and to enhance mosquito control programs throughout the state, we did use a portion of it for basic research on vectors and vector ecology. Results of those studies are provided in the references at the end of this chapter.

Collaborative Research

Entomologists and other scientists at public health departments may choose to collaborate with university or CDC scientists for the study of diseases, vectors, or disease ecology. This is especially useful when the public health entomologist does not have necessary laboratory space, equipment, or expertise to accomplish research objectives. For example, we once performed a statewide tick-borne disease agent survey in Mississippi, but much of the molecular work (DNA extraction and PCR) was performed by collaborators in the rickettsial branch of the CDC in Atlanta.[26] As another example, after Hurricane Katrina, the entomology group in Mississippi relied on the expertise of CDC researchers at Ft. Collins, Colorado, for screening almost 50,000 mosquitoes for arboviral disease agents.[27] Such collaborations are useful for all entities involved and represent wise use of limited financial resources.

> ### SOURCES OF FUNDING FOR HEALTH DEPARTMENTS
> 1. U.S. Centers for Disease Control
> 2. Federal preventive medicine block grants
> 3. Other public money (USDA, etc.)
> 4. Private foundations

Note

1. Information about TDH lab capacity and equipment was provided in 2013 courtesy Dr. Abelardo Moncayo, and may have changed or been updated since that time.

References

1. McCarthy M. Public health research—multidisciplinary, high-benefit, and undervalued. *Innovat Euro J Soc Sci Res.* 2010;23(1):69–77.
2. Quarterman KD. Research in vector control. *Bull World Health Org.* 1963;29 (suppl.):63–68.
3. Eads RB, Irons JV. Current status of public health at the state level. *Am J Public Health Nations Health.* 1951;41:1082–1086.
4. McInnis SJ, Goddard J, Deerman JH, Nations TM, Varnado WC. Insecticide resistance testing of *Culex quinquefasciatus* and *Aedes albopictus* from Mississippi. *J Am Mosq Control Assoc.* 2019;35:147–150.
5. Goddard J. Laboratory assays of various insecticides against bed bugs and their eggs. *J Entomol Sci.* 2013;48:1–5.
6. Goddard J. An annotated list of the ticks (Ixodidae and Argasidae) of Mississippi. *J Vector Ecol.* 2006;31:206–209.
7. Goddard J, Layton MB. A guide to ticks of Mississippi. In: Mississippi Agriculture and Forestry Experiment Station, Mississippi State University, Bulletin Number 1150; 2006:17 pp.
8. Goddard J, Paddock CD. Observations on distribution and seasonal activity of the Gulf Coast tick in Mississippi. *J Med Entomol.* 2005;42:176–179.
9. Goddard J, Piesman J. New records of immature *Ixodes scapularis* from Mississippi. *J Vector Ecol.* 2006;31:421–422.
10. Goddard J, Harrison BA. New, recent, and questionable mosquito records from Mississippi. *J Am Mosq Control Assoc.* 2005;21:10–14.
11. Goddard J, Varnado WC, Harrison BA. An annotated list of the mosquitoes (Diptera: Culicidae) of Mississippi. *J Vector Ecol.* 2010;35(1):213–229.
12. Goddard J, Waggy G, Varnado WC, Harrison BA. Taxonomy and ecology of the pitcher-plant mosquito, *Wyeomyia smithii,* in Mississippi. *Proc Entomol Soc Wash.* 2007;109:684–688.
13. Thorn M, Varnado WC, Goddard J. First record of *Aedes japonicus* in Mississippi. *J Amer Mosq Control Assoc.* 2012;28:43–44.
14. Varnado WC, Goddard J, Harrison BA. New state record of *Culex coronator* Dyar and Knab from Mississippi. *Proc Entomol Soc Wash.* 2005;107:476–477.

15. Goddard J. New records for the phlebotomine sand fly *Lutzomyia shannoni* (Dyar) in Mississippi. *J Mississippi Acad Sci*. 2005;50:195–196.
16. Castellaw AH, Showers J, Goddard J, Chenney EF, Varela-Stokes AS. Detection of vector-borne agents in lone star ticks, *Amblyomma americanum*, from Mississippi. *J Med Entomol*. 2010;47:473–476.
17. Edwards KT, Goddard J, Jones TL, Paddock CD, Varela-Stokes AS. Cattle and the natural history of *Rickettsia parkeri* in Mississippi. *Vector Borne Zoonotic Dis*. 2010;11:485–491.
18. Goddard J, Varela-Stokes A, Nations TM, Portugal JS, Walker A, Varnado WC. Lyme disease agent not detected in deer ticks from Mississippi. *J Mississippi St Med Assoc*. 2018;59:375.
19. Goddard J. American Boutonneuse Fever—a new spotted fever rickettsiosis. *Infect Med*. 2004;21:207–210.
20. Goddard J, Varela-Stokes AS. The discovery and pursuit of American Boutonneuse Fever: a new spotted fever group rickettsia. *Midsouth Entomol*. 2009;2:47–52.
21. Goddard J. Proportion of adult lone star ticks (*Amblyomma americanum*) questing in a tick population. *J Mississippi Acad Sci*. 2009;54:206–209.
22. Goddard J, Goddard J, II. Estimating populations of adult *Ixodes scapularis* in Mississippi using a sequential Bayesian algorithm. *J Med Entomol*. 2008;45:556–562.
23. Goddard J, Goddard J, II. Relative risk of acquiring black-legged ticks, *Ixodes scapularis*, in central Mississippi. *Midsouth Entomol*. 2010;3:97–100.
24. Elnaiem DE, Kelley K, Wright S, et al. Impact of aerial spraying of pyrethrin insecticide on *Culex pipiens* and *Culex tarsalis* (Diptera: Culicidae) abundance and West Nile virus infection rates in an urban/suburban area of Sacramento County, California. *J Med Entomol*. 2008;45(4):751–757.
25. Carney RM, Husted S, Jean C, Glaser C, Kramer V. Efficacy of aerial spraying of mosquito adulticide in reducing incidence of West Nile Virus, California, 2005. *Emerg Infect Dis*. 2008;14(5):747–754.
26. Goddard J, Sumner JW, Nicholson WL, Paddock CD, Shen J, Piesman J. Survey of ticks collected in Mississippi for *Rickettsia, Ehrlichia*, and *Borrelia* species. *J Vector Ecol*. 2003;28:184–189.
27. Goddard J, Varnado WC. Disaster vector control in Mississippi after Hurricane Katrina: lessons learned. *J Am Mosq Control Assoc*. 2020;36:56–60.

chapter eight

Where to Go for Help

State or Local Health Departments

Public health entomological assistance may be found at most state health departments, and in some cases, at local or city health departments. Most state health departments have a state medical entomologist or public health entomologist located in the epidemiology, communicable disease, or environmental health section(s) (see Chapter 3). Some local or city health departments have their own entomology programs. For example, New York City has a Zoonotic and Vector-borne Disease Unit (ZVDU) that conducts disease surveillance, investigations, and education, and provides consultations with human and animal healthcare providers with the goal of preventing and controlling human illness due to zoonotic and vector-borne diseases. A search of the Internet using the keywords "public health entomology," "zoonotic disease," or "vector-borne disease" should identify these state, local, or city programs in your area.

Universities

Many colleges and universities have entomologists. Even if the institution is not large enough to house an entire entomology department, there still may be entomologists in the biology or zoology department(s). Virtually all land-grant institutions in the United States have entomologists. A land-grant college or university is an institution that has been designated by its state legislature or Congress to receive the benefits of the Morrill Acts of 1862 and 1890. These institutions, as set forth in the first Morrill Act, were created to teach agriculture, military tactics, and the mechanic arts, as well as classical studies, so that members of the working classes could obtain a liberal, practical education. There is at least one land-grant college in every state of the union. Over the years, land-grant status has implied several types of federal support. The first Morrill Act provided grants in the form of federal lands to each state for the establishment of a public institution to fulfill the act's provisions. At different times additional monies have been appropriated through legislation such as the second Morrill Act and the Bankhead-Jones Act, although the funding provisions of these acts are no longer in effect. Today, the Nelson Amendment to the Morrill Act provides a permanent annual appropriation of $50,000 per state and territory.

WHERE ENTOMOLOGISTS ARE LOCATED

1. State health departments
2. Land-grant universities (often entomology departments)
3. Biology departments in other colleges and universities
4. Military bases
5. Mosquito abatement districts
6. U.S. Department of Agriculture (Agricultural Research Service)
7. The Centers for Disease Control, Atlanta, GA and Ft. Collins, CO
8. The World Health Organization

A key component of the land-grant system is the agricultural experiment station program created by the Hatch Act of 1887. The Hatch Act authorized direct payment of federal grant funds to each state to establish an agricultural experiment station in connection with the state's land-grant institution. The amount of this appropriation varies from year to year and is determined for each state through a formula based on number of small farms there. A major portion of these federal funds must be matched by the state. In order to disseminate information gleaned from (each state's) experiment stations' research, the Smith-Lever Act of 1914 created a Cooperative Extension Service associated with each U.S. land-grant institution. This act authorized ongoing federal support for extension services, using a formula similar to the Hatch Act's, to determine amount of the appropriation. This act also requires matching funds from states in order to receive the federal monies. The extension service in each state has several entomologists who can be called upon for entomological consultation.

Mosquito Abatement Districts

There are organized mosquito control programs all over the world, varying from centralized to decentralized, disease-specific or generalized, and government-maintained or community-based.[1] Each of these programs operates within its own unique political, economic, social, and technological environment. There are over 700 mosquito abatement districts (MADs) in the United States, which are quasi-governmental agencies with enabling state legislation allowing funding from property taxes or (even) local water bills.[2] Although there have been efforts to eliminate MADs,[3] most have survived, even prospered due to a string of mosquito-borne diseases lately including West Nile virus, chikungunya, and Zika. Most MADs have at least one entomologist on staff, sometimes with an MS or

PhD in entomology. These entomologists are a treasure trove of information about local mosquito and other vector fauna, surveillance data, mosquito infection rates, and the like. Therefore, public health entomologists
should seek to network with state and local MADs and their staff.

USDA Systematic Entomology Laboratory

The U.S. Department of Agriculture (USDA) has an in-house scientific
research agency called the Agricultural Research Service (ARS) whose job
it is to find solutions to agricultural problems affecting Americans every
day from field to table. Currently, the USDA ARS conducts 660 research
projects within 15 national programs, including over 2,000 scientists and
post-docs, with an additional 6,000 other employees, all spread out over
more than 90 research locations. Within the ARS, there is a Systematic
Entomology Laboratory, located at Beltsville, MD, which provides insect
and arachnid identifications for members of public health and academia.
Public health entomologists wishing to use this service should read and
follow the submission guidelines provided on their website.[4]

Centers for Disease Control and Prevention

The CDC has several centers specializing in different areas of human
health and disease prevention. Two of these contain entomology-related
programs—the National Center for Emerging and Zoonotic Infectious
Diseases (NCEZID) and the Center for Global Health (CGH), which were
reorganized in 2010 under the CDC's Office of Infectious Diseases to provide leadership, expertise, and service in laboratory and epidemiological
science, bioterrorism preparedness, applied research, disease surveillance, and outbreak response for infectious diseases. NCEZID has two
divisions specifically dealing with entomological issues—the Division of
Vector-Borne Diseases and the Division of High Consequence Pathogens
and Pathology. The Vector-Borne Infectious Diseases Division has an
Arboviral Diseases Branch, a Bacterial Diseases Branch, a Rickettsial
Zoonoses Branch, and a Dengue Branch, emphasizing dengue, West Nile,
and other encephalitis viruses, Lyme disease, plague, tularemia, Rocky
Mountain spotted fever, ehrlichiosis, and anaplasmosis. The Division of
High Consequence Pathogens and Pathology includes the Viral Special
Pathogens Branch, which deals with arthropod-borne hemorrhagic
fevers such as Crimean-Congo hemorrhagic fever and Kyasanur forest
disease, and the Infectious Diseases Pathology Branch, which provides
consultation and diagnostic support to all of the other branches and divisions within the CDC. The Division of Parasitic Diseases and Malaria
within CGH has a Malaria Branch, a Parasitic Diseases Branch, and an

Entomology Branch that provide consultation and conduct research on such things as malaria, leishmaniasis, and Chagas disease. For questions about laboratory diagnosis of parasitic diseases, the CDC DPDx website is a helpful resource.[5] In addition to providing information on the latest diagnostic techniques, DPDx includes dozens of photographs and figures of various parasites occurring worldwide.

World Health Organization

The World Health Organization (WHO) is a specialized agency operating under the auspices of the United Nations. Shortly after the United Nations was organized in 1945, diplomats recognized the need for an international health agency and the WHO constitution was approved on April 7, 1948.[6] Currently, 193 nations, plus several associate members such as the Vatican, are members of the WHO. Organizationally, there are 147 country offices, 6 regional offices, and a main headquarters located in Geneva, Switzerland. More than 8,000 public health experts from various nationalities work for the WHO, and because the organization deals with public health policies, guidelines, and research, these employees are doctors, epidemiologists, scientists, managers, administrators, and other professionals. Entomology-related programs and research at the WHO include the Global Malaria Program (GMP), the Neglected Tropical Diseases Program (NTD), the Global Program to Eliminate Lymphatic Filariasis (GPELF), and the Special Program for Research and Training in Tropical Diseases (TDR), just to name a few. The Global Malaria Program (GMP) is well funded with over 100 partners, such as the Bill and Melinda Gates Foundation, and intends to stop new cases of malaria by 2025, a goal not likely to be reached. However, the GMP is aggressively promoting mosquito control, use of bed nets for personal protection against mosquito biting, and early case detection and treatment regimes that hopefully will reduce malaria incidence worldwide. WHO's neglected tropical diseases (NTD) effort includes several vector-borne diseases such as leishmaniasis and dengue, which are responsible for millions of cases of morbidity and mortality each year. The human filariasis elimination program (GPELF) was launched by WHO in 2000 and has been quite successful, having provided treatment for over 2.5 billion people worldwide (about 50% of the at-risk population) with the goal of eliminating the disease by 2030.[7,8] The research and training program in tropical diseases (TDR) was founded in 1975 to address research and training in neglected tropical diseases and development of new treatments and diagnostic tools for them. TDR prioritizes research into NTDs by assessing disease burden, incidence, current knowledge of disease, and cost-effectiveness of various proposed interventions.

There are other public-private partnerships operated through the WHO that address entomological issues such as the UN/WHO Program against African Trypanosomiasis (PAAT), the African Program for Onchocerciasis Control (APOC), and the Onchocerciasis Elimination Program of the Americas (OEPA). In addition, the WHO helps develop and provide outstanding entomological training materials available online, such as the "Entomology in Public Health Practice" chapter of a textbook on public health published in cooperation with the WHO India office.[9]

References

1. Impoinvil DE, Ahmad S, Troyo A, et al. Comparison of mosquito control programs in seven urban sites in Africa, the Middle East, and the Americas. *Health Policy.* 2007;83:196–212.
2. Kerzee R. The whats and whys of mosquito abatement districts. In: Midwest Grows Green Newslewtter, The IPM Institute of North America, Madison, WI; 2018;June Issue:2 pp.
3. Ruthhart B. Push to eliminate mosquito-fighting layer of government stirs passion on both sides. *Chicago Tribune.* June 28, 2011:13–15.
4. USDA. Systematic Entomology Laboratory, Beltsville, MD; 2021. www.ars.usda.gov/northeast-area/beltsville-md-barc/beltsville-agricultural-research-center/systematic-entomology-laboratory/.
5. CDC. Division of parasitic diseases diagnosis website. In: CDC, DPD, Laboratory Diagnostic Assistance; 2011. www.dpd.cdc.gov/dpdx/HTML/Contactus.htm.
6. WHO. Working for health: an introduction to the World Health Organization. In: WHO; 2007. www.who.int/about/brochure_en.pdf.
7. WHO. Global programme to eliminate lymphatic filariasis. *Wkly Epidemiol Rec.* 2010;85:365–372.
8. WHO. Lymphatic filariasis: reporting continued progress towards elimination as a public health problem. In: World Health Organization, Geneva, Switzerland;2021. www.who.int/news/item/29-10-2020-lymphatic-filariasis-reporting-continued-progress-towards-elimination-as-a-public-health-problem.
9. Tilak R. Entomology in public health practice. In: Bhalwar R, ed. *Textbook on Public health and Community Medicine.* Pune, India: Department of Community Medicine, Armed Forces Medical College, in association with WHO/India; 2009:903–975.

section two

Some Primary Pests and Conditions of Public Health Importance

chapter nine

Mosquitoes

Importance and Physical Description

Mosquitoes are by far the most important bloodsucking arthropods worldwide, causing much annoyance and disease in humans, other mammals, and birds. About 3,500 species of mosquitoes have been described, but relatively few of them are significant vectors of human diseases; however, the mosquito-transmitted disease burden worldwide is quite severe.

Mosquitoes are flies in the order Diptera (Figure 9.1) that can be distinguished from other flies by several characteristics. They have long,

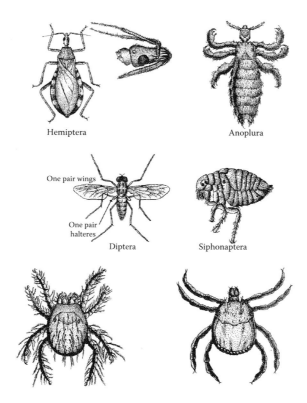

Hemiptera   Anoplura

One pair wings

One pair halteres

Diptera   Siphonaptera

Figure 9.1 Various arthropod orders of medical importance.

Source: From the U.S. Navy

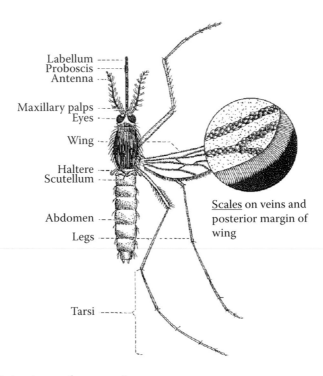

Labellum
Proboscis
Antenna

Maxillary palps
Eyes

Wing

Haltere
Scutellum

Abdomen

Legs

Scales on veins and
posterior margin of
wing

Tarsi

Figure 9.2 Anatomy of a mosquito.

Source: From the U.S. Navy

15-segmented antennae, a long proboscis for bloodsucking, and scales on the wing fringes and veins (Figure 9.2). Scale patterns on the "back" (scutum) and legs of mosquitoes can sometimes be used to identify the species. The mosquito head is rounded, bearing large compound eyes that almost meet. Males can usually be distinguished from females by their bushy antennae. Female *Anopheles* mosquitoes (the vectors of malaria) have palpi as long as the proboscis; other groups do not have this characteristic.

Distribution

Mosquitoes breed in water and occur worldwide wherever conditions are favorable for their development. Contrary to what one may think, mosquitoes are particularly worrisome in very cold areas such as Alaska and Canada. Going outdoors in certain areas of Alaska and Canada during summer months is almost unbearable due to snowmelt mosquitoes.

Figure 9.3 Adult female mosquito feeding.
Source: From the Centers for Disease Control

Impact on Human Health

Malaria. Due to their bloodsucking habit (Figure 9.3) mosquitoes may pick up and later transmit disease agents. Malaria has historically been one of the most significant human health threats in the entire world (Figure 9.4). To this day it remains prevalent in many areas of the world (especially so in Africa), resulting in staggering case numbers and deaths (Figure 9.5). Recent estimates put the case numbers at about 230 million cases per year with 500,000 deaths.[1,2] Previously, human malaria was thought to be caused by any one of four species of microscopic protozoan parasites in the genus *Plasmodium*—*Plasmodium vivax, Plasmodium ovale, Plasmodium malariae*, and *Plasmodium falciparum*, although we now know that a fifth species may cause the disease—*Plasmodium knowlesi*.[3–5] The disease ecology of malaria is complicated as not all species of *Plasmodium* occur in all places, nor do they produce identical disease syndromes. Further, climate change may be resulting in the spread of malaria to new areas; for example, malaria has recently spread into highland regions of East Africa where it did not exist previously. This spread presumably occurred because of warmer and wetter weather, resulting in high rates of illness and death because the disease was introduced into a largely susceptible population.[6,7] Complicating the malaria situation, mosquito vectors of malaria are becoming resistant to many of the pesticides being used to control them, as well as to the insecticides used in insecticide treated bed nets

Figure 9.4 Woman with malaria in India desperately awaiting medical treatment, 1955.

Source: From the World Health Organization

(ITNs), and (lastly) the malaria parasites themselves are becoming resistant to drugs used to prevent the disease, even artemisinin-based combination therapy (ACT) drugs.

Yellow fever. Yellow fever (YF) is probably the most lethal of all arboviruses and has had a devastating effect on human social development.

Figure 9.5 Approximate geographic distribution of malaria.

The causative agent, a flavivirus, now primarily occurs in Africa and South and Central America (Figure 9.6). Forty-seven countries in Africa and South America are endemic for YF.[8] Mild cases may be characterized by fever, headache, generalized aches and pains, and nausea. Persons with severe YF may exhibit high fever, headache, dizziness, muscular pain, jaundice, hemorrhagic symptoms, and profuse vomiting of brown or black material, often resulting in collapse and death. Even though there is an effective vaccine, the WHO estimates 84,000–170,000 severe cases occur each year with 29,000–60,000 deaths.[8]

Dengue fever. Dengue is caused by a virus in the family Togaviridae, and is responsible for widespread morbidity (breakbone fever) and some mortality [dengue hemorrhagic fever (DHF)] in much of the tropics and subtropics each year (Figure 9.7). Most dengue infections result in relatively mild illness characterized by fever, headache, myalgia, rash, nausea, and vomiting. However, DHF may be severe, causing petechiae, purpura, mild gum bleeding, nosebleeds, gastrointestinal bleeding, and dengue shock syndrome. Worldwide, as many as 400 million cases of dengue occur annually, and several hundred thousand cases of DHF.[9] Although the disease literally "knocks on the door" of the continental United States each year, it infrequently becomes established here. However, in 2009, there was an outbreak of locally acquired dengue in the Florida Keys.[10]

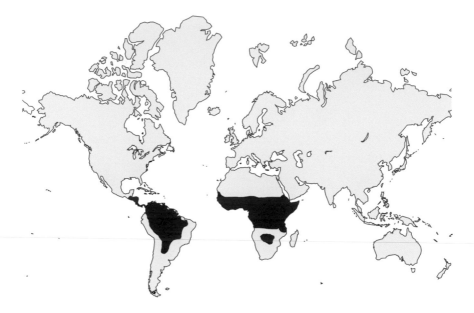

Figure 9.6 Approximate geographic distribution of yellow fever.

Figure 9.7 Approximate geographic distribution of dengue.

Lymphatic filariasis. Several species of nematode worms may cause lymphatic filariasis, an important human mosquito-borne disease occurring in much of the world (Figure 9.8). Malayan filariasis, caused by *Brugia malayi*, is mostly confined to Southeast Asia, and the Bancroftian form, *Wucheria bancrofti*, is prevalent over much of the tropical world. In 2000, the WHO estimated that 120 million people were infected with Bancroftian or Brugian filariasis, with an additional 1.34 billion persons at risk.[11] That number is now significantly lower due to mass drug administration using ivermectin, diethylcarbamazine, and other compounds.[12] In the Western Hemisphere, 80% of lymphatic filariasis occurs in Haiti, likely imported from Africa with the slave trade.[13] Human filariasis is transmitted solely by mosquitoes, and there is no multiplication of the parasite, only development, in the mosquito vector. In addition, the adult worms may live up to 10 years in humans.[14]

Encephalitis viruses. In temperate North America, the worst mosquito-borne diseases are probably the encephalitides. Certainly not all cases of encephalitis are mosquito caused (enteroviruses and other agents are often involved), but mosquito-borne encephalitis has the potential to cause serious morbidity and mortality covering widespread geographic areas each year. For example, West Nile virus (WNV) has become endemic in the United States and has ongoing potential for seasonal epidemics at the local, regional, and national level.[15] There were 2,647 cases of WNV reported in the United States in 2018.[15,16] In addition to the actual disease

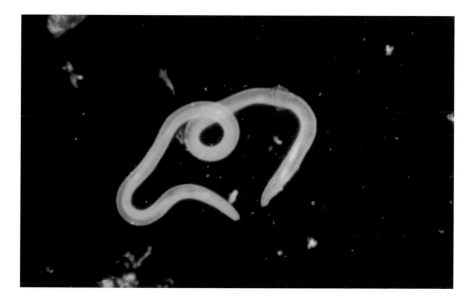

Figure 9.8 Brugia malayi, one agent of lymphatic filariasis, FA stain, 125X.

Source: From the Centers for Disease Control, photo by A.V. Sulzer

burden, there are significant economic costs associated with WNV and related encephalitis viruses. For example, the estimated cost of WNV in one state (Louisiana) in 2002–2003 was $20.1 million, with almost half of that being cost of the public health response.[17] WNV cases are generally divided into two groups—West Nile fever and the more serious WNV neuroinvasive disease. During 2009–2018, approximately 20,000 cases of WNV neuroinvasive disease were reported from all states, DC, and Puerto Rico, with the highest incidence in the North Central region of the United States (Figure 9.9).[15] Another serious encephalitis virus is eastern equine encephalitis (EEE), which caused an outbreak in 7 states during 2019 where 34 people were sickened with 12 deaths.[18] EEE is known for producing a high mortality rate (30–75%) and severe permanent neuro-logic sequelae in people who survive the disease.[19]

Rift Valley fever. Rift Valley fever (RVF), occurring throughout sub-Saharan Africa and Egypt, is a Phlebovirus in the family Bunyaviridae. It causes abortions in sheep, cows, and goats and heavy mortality of lambs and calves. In humans, RVF may produce fever, myalgias, encephalitis, hemorrhage, or retinitis. Permanent visual impairment has been reported. Huge epidemics occasionally occur. For example, in 1950–1951 there were an estimated 100,000 sheep and cattle cases and at least 20,000 human cases. Outbreaks continue; in 1998, there were 27,500 human cases and 170 deaths from RVF in Kenya.[20] Although RVF may be acquired by handling infective material of animal origin during necropsy or butchering, it is primarily a vector-borne zoonotic pathogen transmitted by Aedes mos-quitoes.[21] Recently, the distribution of RVF seems to be expanding.

Ross River virus disease. Ross River (RR) disease, also called epidemic polyarthritis, occurs throughout most of Australia and occasionally New Guinea, Fiji, Tonga, and the Cook Islands. It causes fever, headache, fatigue, rash, and—most notably—arthritis in the wrist, knees, ankles, and small

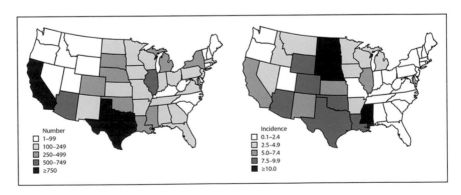

Figure 9.9 West Nile virus incidence map, 2009–2018.

Source: From the Centers for Disease Control

joints of the extremities. The disease is not fatal but may be debilitating, with symptoms occurring for weeks or months. RR disease is the most common arboviral disease in Australia with more than 5,000 cases annually.[22] During 1997, there were 6,683 cases reported.[23] Peak incidence occurs from January through March, when mosquito vectors are most abundant. Research has indicated that most likely kangaroos and wallabies are natural hosts for RR virus. There are several mosquito vectors of RR virus in Australia, but particularly *Culex annulirostris* inland and Aedes vigilax and Ae. camptorhynchus in northern and southern coastal areas, respectively.

Chikungunya fever. Chikungunya (CHIK) is a mosquito-transmitted Alphavirus which is not usually fatal but can cause severe fevers, headaches, fatigue, nausea, and muscle and joint pains.[24,25] CHIK may cause excruciatingly painful swelling of the joints in fingers, wrists, back, and ankles. The virus was first isolated during an epidemic in Tanzania where the Swahili word Chikungunya means "that which bends up," referring to the position patients assume while suffering severe joint pains.[25] The virus may be transmitted by *Aedes aegypti, Ae. albopictus, and Ae. polynesiensis* (Polynesian islands). The geographic distribution of CHIK has historically included most of Sub-Saharan Africa, India, Southeast Asia, Indonesia, and the Philippines, although the disease is increasing both in incidence and geographic range. There were at least 300,000 cases on Reunion Island in the Indian Ocean during 2005–2006. India suffered an explosive outbreak in 2006 with more than 1.25 million cases. CHIK was found in Italy in 2007.[24,26] During 2013–2014, there were thousands of cases reported in the Caribbean and Central, and South America.[27] In 2015, there were 896 cases of travel-related CHIK in the United States and one locally acquired case.[28] One of the mosquito vectors of CHIK, the Asian tiger mosquito, *Ae. albopictus*, is extremely abundant in the southern United States, raising fears of widespread outbreaks should local mosquitoes become infected.[26]

Zika virus. Zika virus (ZIKV), a rather mild mosquito-borne human disease, is transmitted by *Aedes aegypti, Ae. albopictus, Ae. polynesiensis*, and *Ae. hensilli*. The virus is a RNA virus in the family *Flaviviridae*, which includes several other viruses of clinical importance. ZIKV is closely related to Spondweni virus, the only other member of its group. Genetic analysis shows that ZIKV can be classified into distinct African and Asian lineages, both emerging from East Africa during the late 1800s or early 1900s.[29] Infection is asymptomatic in 80% of cases, making reporting and (necessary) public health interventions difficult. People of all ages are susceptible to ZIKV (4 days–76 years), with a slight preponderance of cases in females.[29,30] If symptoms occur, they are typically temporary, self-limiting, and nonspecific. Commonly reported symptoms include rash, fever, arthralgia, myalgia, fatigue, headache, and conjunctivitis.[31] Rash, a prominent feature, is maculopapular and pruritic, and usually begins proximally then spreading to the extremities with spontaneous resolution within 1–4 days of onset.[29] There may be a low grade fever

(37–38°C). Symptoms generally resolve within 2 weeks and reports of longer persistence are rare. The most important medical issues associated with Zika are the severe clinical sequelae Guillian-Barré syndrome and microcephaly (Figure 9.10).[30] During the 2015 outbreak in Brazil, reports of infants born with microcephaly greatly increased (>3,800 cases; 20 cases/10,000 live births vs. 0.5/10,000 live births in previous years).[29] During 2015–2017, there were millions of cases of ZIKV throughout the Americas and approximately

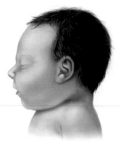

Baby with Typical Head Size

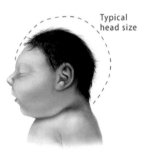

Baby with Microcephaly

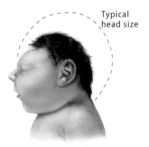

Baby with
Severe Microcephaly

Figure 9.10 Microcephaly due to Zika virus.

Source: From the Centers for Disease Control

275 locally acquired cases occurred in the United States.[31,32] All cases in the United States during 2016–2017 were associated with *Ae. aegypti* mosquitoes and not *Ae. albopictus* (which is fortunate because *Ae. albopictus* is much more widespread). Cases declined rapidly after 2017, but may resurface again after a few years since arboviruses are often cyclical.

MAJOR MOSQUITO-BORNE DISEASES WORLDWIDE

- Malaria
- Dengue
- Yellow fever
- Lymphatic filariasis (several types)
- Encephalitis viruses (e.g., West Nile, eastern equine, LaCrosse, Japanese encephalitis)
- Other arboviruses (e.g., Chikungunya, Zika, Ross River, etc.)

Prevention and Control

Prevention and control of mosquitoes involves source reduction, larviciding, and adulticiding. Large area spraying for mosquitoes is usually done with truck-mounted ultra-low volume (ULV) machines (Figure 9.11). Source

Figure 9.11 Broad area mosquito spraying is usually done by truck-mounted ULV machines.

Source: Photo copyright 2020 by Jerome Goddard, Ph.D.

reduction involves physical/mechanical measures to eliminate mosquito breeding sites including removal of artificial containers such as old tires, filling in low spots in fields and yards, cleaning out ditches so they flow better (Figure 9.12), and/or draining woodland pools. Trenching swamps and other wetlands was historically used for mosquito control, but has since been abandoned due to environmental concerns. Interestingly, an

Figure 9.12 Cleaning out ditches can help prevent mosquito breeding.

Source: Photo copyright 2020 by Jerome Goddard, Ph.D.

over-looked mosquito prevention tool may be controlled burning of veg-etation, presumably by killing mosquito eggs.[33]

IPM and Alternative Control Methods

Integrated pest management (IPM) is a strategy to minimize negative environmental impacts by drawing upon a variety of pest control mea-sures, including both chemical and nonchemical measures. The objec-tives of classic IPM are to: (1) prevent unacceptable levels of pest damage, (2) minimize the risk to people, property, infrastructure, natural resources, and the environment, and (3) reduce the evolution of pest resistance to pes-ticides and other pest management practices.[34] IPM for mosquitoes, some-times referred to as Integrated Mosquito Management (IMM), mainly consists of eliminating breeding sites, emphasis on larviciding instead of spraying for adult mosquitoes, precision-targeting (indoors) of residual insecticides, and surveillance-based adulticiding (outdoors) instead of routine adulticiding. For example, a recent study found no significant dif-ferences between numbers of *Anopheles* mosquitoes landing on humans in huts with 50% of their walls sprayed with deltamethrin insecticide as opposed to 100% of indoor walls sprayed, demonstrating that the extent of indoor spraying can be reduced and still provide protection.[35] Practical nonchemical measures for mosquito avoidance include limiting outdoor activity after dark and avoiding known mosquito-infested areas (e.g., swamps, marshes) during the peak mosquito season. In addition, people who need to be outdoors after dark in mosquito season should wear long sleeves and long pants. Probably one of the most basic and effective sani-tation measures to limit mosquito–human contact is that of screen wire windows and doors. Screens may consist of various metals or plastic and are ordinarily 16 × 16 × 20 mesh. They should be tight fitting over window openings. Screen doors should be hung so that they open outward. Being such a basic protection measure, screens are sometimes overlooked; how-ever, their importance cannot be overemphasized. During the mosquito season, people may choose to use protective head gear or jackets made of netting when outdoors in heavily infested areas. Also, those camping may sleep under mosquito nets for protection from mosquitoes. This becomes mandatory on safaris or other trips to the tropics or subtropics for protec-tion against disease-carrying mosquitoes (Figure 9.13). Ordinarily, mos-quito netting is made of cotton or nylon with 23 to 26 meshes per inch. Netting should not be allowed to lie loosely on the head or body, because mosquitoes can feed through the net wherever it touches the skin; there-fore, construction of a crude frame may help keep the net away from the body. Insecticide-treated nets (ITNs) are a mainstay for malaria and fila-riasis prevention in Africa, southeast Asia, and South America, and they are widely promoted by the World Health Organization and charitable

Figure 9.13 Mosquito bed nets are effective in preventing malaria.

Source: From the Centers for Disease Control, photo by B.K. Kapella

groups such as the Bill and Melinda Gates Foundation. These nets are usually treated with pyrethroid insecticides such as permethrin or deltamethrin and may remain effective for months or even a year or two if not rinsed or washed. One study in Papua New Guinea showed that bites from mosquitoes ranged from 6.4 to 61.3 bites per person per day before bed net distribution, and from 1.1 to 9.4 bites for 11 months after distribution (P<0.001).[36] However, a recent worrisome trend is insecticide resistance to the pyrethroids in ITNs, leading to loss of efficacy,[37] although their use is still considered very effective.[38]

References

1. Anonymous. Is malaria elimination within reach? *Lancet Infect Dis.* 2017;17:461.
2. WHO. World malaria report, 2020. In: World Health Organization, Geneva, Switzerland; 2020. www.who.int/teams/global-malaria-programme/reports/world-malaria-report-2020.
3. Collins WE, Barnwell JW. *Plasmodium knowlesi*: finally being recognized. *J Infect Dis.* 2009;199(8):1107–1108.
4. Indra V. *Plasmodium knowlesi* in humans: a review on the role of its vectors in Malaysia. *Trop Biomed.* 2010;27(1):1–12.
5. Kantele A, Jokiranta S. *Plasmodium knowlesi* – the fifth species causing human malaria. *Duodecim.* 2010;126(4):427–434.

6. Lafferty KD. The ecology of climate change and infectious diseases. *Ecology*. 2009;90:888–900.
7. Shuman EK. Global climate change and infectious diseases. *N Engl J Med*. 2010;362:1061–1063.
8. WHO. Yellow fever fact sheet; 2020. In: www.who.int/news-room/fact-sheets/detail/yellow-fever.
9. Bhatt S, Gething PW, Brady OJ, et al. The global distribution and burden of dengue. *Nature*. 2013;496(7446):504–507.
10. Radke EG, Gregory CJ, Kintzinger KW, et al. Dengue outbreak in Key West, Florida, USA, 2009. *Emer Infect Dis*. 2012;18:135–137.
11. WHO. Global programme to eliminate lymphatic filariasis. *Wkly Epidemiol Rec*. 2010;85:365–372.
12. WHO. GPELF progress report 2019. In: World Health Organization, Geneva, Switzerland; 2020. www.who.int/publications/i/item/who-wer9543.
13. Roberts L. Relief among the rubble. *Science*. 2010;327.
14. Laurence BR. The global distribution of bancroftian filariasis. *Parasitol Today*. 1989;5:260–264.
15. CDC. Surveillance for West Nile virus disease—United States, 2009–2018. In: CDC, MMWR, Surveillance Summaries. 2021;70(1):1–16.
16. CDC. Summary of notifiable diseases. In: National Notifiable Diseases Surveillance System (NNDSS); 2018. wwwn.cdc.gov/nnds/infections-tables-html.
17. Zohrabian A, Meltzer MI, Ratard R, et al. West Nile virus economic impact, Louisiana, 2002. *Emerging Infect Dis*. 2004;10(10):1736–1744.
18. CDC. Notes from the field: multistate outbreak of eastern equine encephalitis virus—United States, 2019. In: MMWR. 2020;69:50–51.
19. Armstrong PM, Andreadis TG. Eastern equine encephalitis virus—old enemy, new threat. *N Eng J Med*. 2013;368:1670–1673.
20. Woods CW, Karpati AM, Grein T, et al. An outbreak of Rift Valley fever in northwestern Kenya. *Emerg Infect Dis*. 2002;8:138–142.
21. CDC. Rift Valley Fever outbreak—Kenya, November 2006-January 2007. In: CDC, MMWR. 2007;56:73–76.
22. Musso D, Rodriguez-Morales AJ, Levi JE, Cao-Lormeau V, Gubler DJ. Unexpected outbreaks of arboviral infections: lessons learned from the Pacific and tropical America. *The Lancet*. 2018; doi: 10.1016/S1473-3099(18)30269-X.
23. Anonymous. Ross River virus. In: New South Wales Dept. Health, Fact Sheet, Sydney, Australia; 2002:1.
24. Enserink M. Tropical disease follows mosquitoes to Europe. *Science (News Focus)*. 2007;317:1485.
25. Weaver SC, Smith DW. Alphavirus infections. In: Guerrant RL, Walker DH, Weller PF, eds. *Tropical Infectious Diseases*. 3rd ed. London: Saunders (Elsevier); 2011:519–524.
26. Enserink M. Chikungunya: no longer a Third World disease. *Science (News Focus)*. 2007;318:1860–1861.
27. Leparc-Goffart I, Nougairede A, Cassadou S, Prat C, de Lamballerie X. Chikungunya in the Americas. *Lancet*. 2014;383:514.
28. CDC. Summary of notifiable infectious diseases and conditions—United States, 2015. In: CDC, MMWR. 2017;64(53):1–144.
29. Plourde AR, Bloch EM. A literature review of Zika virus. *Emerg Infect Dis*. 2016;22(7):1185–1192.

30. Petersen LR, Jamieson DJ, Honein MA. Zika virus. *N Engl J Med.* 2016;375(3):294–295.
31. Gatherer D, Kohl A. Zika virus: a previously slow pandemic spreads rapidly through the Americas. *J Gen Virol.* 2016;97:269–273.
32. CDC. Cumulative Zika virus disease case counts in the United States, 2015–2017. In: CDC website; 2017. www.cdc.gov/zika/reporting/case-counts.html.
33. Scasta JD. Fire and parasites: an under-recognized form of anthropogenic land use change and mecahnism of disease exposure. *EcoHealth.* 2015;doi: 10.1007/s10393-015-1024-5.
34. USDA. A national roadmap for integrated pest management. In: The Federal IPM Coordinating Committee Report; 2018. www.ars.usda.gov/ARSUserFiles/OPMP/IPM%20Road%20Map%20FINAL.pdf.
35. Tainchum K, Bangs MJ, Sathantriphop S, Chareonviriyaphap T. Effect of different wall surface coverage with deltamethrin-treated netting on the reduction of indoor-biting *Anopheles* mosquitoes *J Med Entomol.* 2021;doi: 10.1093/jme/tjab095.
36. Reimer LJ, Thomsen EK, Tisch DJ, et al. Insecticidal bed nets and filariasis transmission in Papua New Guinea. *N Engl J Med.* 2013;369(8):745–753.
37. N'Guessan R, Corbet V, Akogbeto M, Rowland ME. Reduced efficacy of insecticide-treated nets and indoor residual spraying for malaria control in pyrethroid resistance area, Benin. *Emerg Infect Dis.* 2007;13:199–206.
38. Tokponnon FT, Ogouyémi AH, Sissinto Y, et al. Impact of long-lasting, insecticidal nets on anaemia and prevalence of *Plasmodium falciparum* among children under five years in areas with highly resistant malaria vectors. *Malaria J.* 2014;13(76).

chapter ten

Ticks

Importance and Physical Description

Ticks are bloodsucking ectoparasites (Figure 10.1) that are efficient vectors of several different types of disease agents, such as bacteria, viruses, rickettsiae, and protozoans. In fact, they are second only to mosquitoes as arthropod vectors of human disease. In addition, tick bites may cause a variety of acute and chronic skin lesions, as well as a paralysis syndrome resulting from salivary toxins injected during feeding.

Ticks are flat, disc-shaped organisms with one apparent body region and eight legs. In reality, ticks are simply large mites. There are three families of ticks recognized in the world today: (1) Ixodidae (hard ticks), (2) Argasidae (soft ticks), and (3) Nuttalliellidae (a small, curious, little-known family with some characteristics of both hard and soft ticks). The terms *hard* and *soft* refer to the presence of a dorsal scutum or "plate" in the Ixodidae, which is absent in the Argasidae (Figure 10.2). Hard ticks display sexual dimorphism, wherein males and females look obviously different (Figure 10.1), and the blood-fed females are capable of enormous expansion before egg laying (Figure 10.3). The resulting tick larvae are six-legged, instead of eight (Figures 10.4 and 10.5). Hard tick mouthparts are anteriorly attached and visible in the dorsal view. If eyes are present, they are located dorsally on the sides of the scutum. Soft ticks are often globular in appearance with a granular, mammillated, or tuberculated surface (Figure 10.6). They do not display sexual dimorphism and the mouthparts cannot be seen in the dorsal view.

Figure 10.1 Adult female and male *Dermacentor andersoni*, typical hard ticks.

Source: Courtesy of Dr. Blake Layton, Mississippi State University Extension Service

197

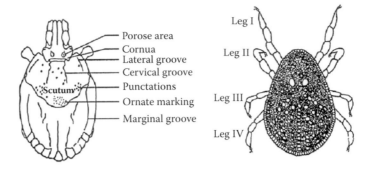

Figure 10.2 Hard tick (left) versus soft tick (right) showing scutum, a diagnostic character for hard ticks.

Source: Compiled from the USDA Agriculture Handbook No. 485

Figure 10.3 Female hard tick fully engorged.

Source: Courtesy of Dr. Blake Layton, Mississippi State University Extension Service

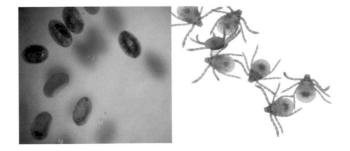

Figure. 10.4 Hard tick eggs (left) and larvae (right).

Source: Photo copyright 2011 by Jerome Goddard, Ph.D.

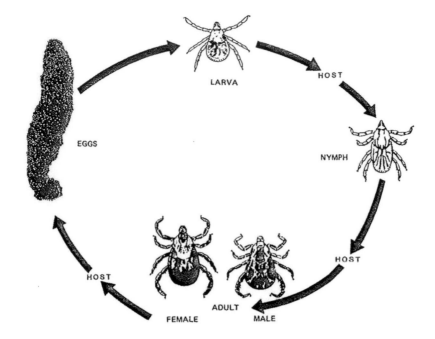

Figure 10.5 Tick life cycle.

Source: Figure courtesy Tennessee Valley Authority, Knoxville, TN

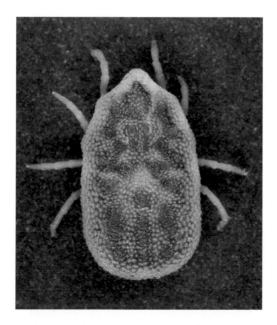

Figure 10.6 Example of a soft tick showing mamillated surface.

Source: From the Centers for Disease Control, photo Dr. William Nicholson

Distribution

Hard ticks and soft ticks occur almost worldwide in different habitats. In general, hard ticks occur in brushy, wooded, or weedy areas containing numerous deer, cattle, dogs, small mammals, or other hosts. Soft ticks are generally found in animal burrows or dens, bat caves, dilapidated or poor-quality human dwellings (huts, cabins, etc.), or animal-rearing shelters. Many soft tick species thrive in hot and dry conditions, whereas hard ticks are more sensitive to desiccation (the genus *Hyalomma* may be an exception), and therefore usually found in areas protected from high temperatures, low humidities, and constant breezes.

Impact on Human Health

Ticks may transmit a wide variety of human disease agents, and are ranked here roughly in order of clinical relevance and frequency seen in a clinic.

Lyme disease. *Lyme disease* or *Lyme borreliosis* (LB), caused by the spirochete *Borreliella burgdorferi*, is a systemic tickborne illness with many clinical manifestations that occurs over much of the world in temperate zones (Figure 10.7). NOTE: Recently, researchers have moved all LB-producing bacterial species into the genus *Borreliella*, leaving the tick-borne relapsing fever-producing species in the genus *Borrelia*, causing some confusion (see textbox).[1,2] Although rarely fatal, Lyme borreliosis may be long and debilitating with cardiac, neurologic, and joint involvement. The most recognizable clinical sign of LB is an expanding bull's eye rash called erythema migrans (Figure 10.8). The number of reported LB cases in the United States continues on an upward trend. There were 33,666 cases reported to the CDC in 2018,[3,4] but an estimate based on insurance claims data is 476,000 cases per year.[5] In the United States, the vast majority of cases are from the northeastern and north-central states (Figure 10.7) and are transmitted by the deer tick, *Ixodes scapularis*. Other Lyme-like illnesses have been described in the medical literature and this causes confusion. In the southern United States, for example, the Centers for Disease Control often labels these Lyme-like illnesses as Southern Tick-associated Rash Illness (STARI) (Figure 10.9). Cases of STARI may be due to allergic reactions to tick saliva or other (as yet) unknown causes.[6] Many physicians still persist in diagnosing such lesions resulting from tick bites as the erythema migrans lesion of Lyme borreliosis.[6] However, there are reports of erythema migrans in humans without any evidence of infection with the LB agent.[7] Diagnosis of Lyme borreliosis is generally based upon clinical presentation which can be more or less accurately confirmed (depends on who you ask) by a two-step procedure: a sensitive enzyme-linked immunosorbent assay followed by immunoblot (IgM and IgG) if reactive. There have been proposals to change this recommended 2-tier algorithm to one consisting of two different types of ELISA tests (and eliminate the Western blot).[8] This approach would make the tests easier to perform, reduce subjectivity in interpreting Western blots, and be cheaper. It should be noted that there are a number of tests for LB which are unreliable and not recommended, such as various tests of bodily fluids, polymerase chain reaction (PCR) of urine, and lymphocyte transformation tests.[9-11]

Figure 10.7 Approximate geographic distribution of Lyme borreliosis.

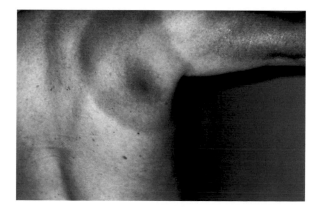

Figure 10.8 Erythema migrans lesion of Lyme borreliosis.

Source: Photo courtesy the Centers for Disease Control

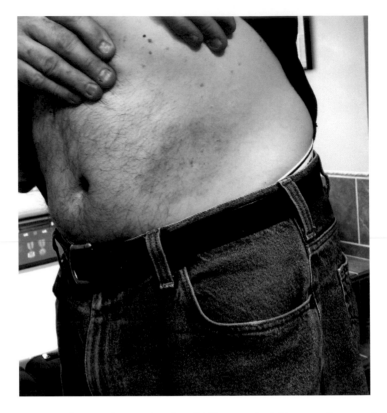

Figure 10.9 Southern tick-associated rash illness, mimicking erythema migrans.

Source: Photo copyright 2016 by Jerome Goddard, Ph.D.

Spotted fever group rickettsioses. Ticks may transmit a wide variety of rickettsial organisms, classified by scientists into several distinct groups. One of the main groups, the spotted fever group (SFG), contains rickettsial species related to the agent of Rocky Mountain spotted fever (RMSF), *Rickettsia rickettsii*. But there are many other rickettsial species in the SFG (Figure 10.10); it contains at least 8 disease agents and 15 others with low or no pathogenicity to humans. RMSF is the most frequently reported rickettsial disease in the United States, with several thousand cases reported each year. In 2018, there were 5,544 cases of spotted fever rickettsiosis reported in the United States.[4] At the time of initial presentation, there is often the classic triad of RMSF: fever, headache, and rash (Figure 10.11). Other characteristics are malaise, chills, myalgias, and gastrointestinal symptoms. Sometimes RMSF leads to coma and death, and the mortality rate is about 5% even with treatment.

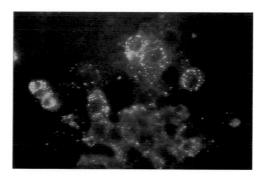

Figure 10.10 Spotted fever group rickettsiae in cell culture.

Source: Figure courtesy Dr. Andrea Varela-Stokes, Mississippi State University

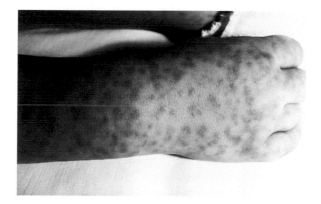

Figure 10.11 Rocky Mountain spotted fever rash in a child.

Source: Photo courtesy the Centers for Disease Control

Ehrlichiosis and anaplasmosis. *Ehrlichia* and *Anaplasma* organisms may be transmitted by ticks as well. They are rickettsia-like bacteria that primarily infect circulating leukocytes. The most common of them, *Ehrlichia chaffeensis*, the causative agent of *human monocytic ehrlichiosis* (HME), occurs mostly in the central and southern United States, and infects mononuclear phagocytes in blood and tissues.[12] There were 1,799 cases of HME in the United States in 2018.[4] A new species of *Ehrlichia* causing human illness in Minnesota and Wisconsin has recently been recognized.[13] Another, *Anaplasma* (formerly *Ehrlichia*) *phagocytophilum*, infects granulocytes and causes *human granulocytic anaplasmosis* (HGA); it is mostly reported from the upper Midwest and northeastern United States. There were 4,008 cases of HGA in the United States in 2018.[4]

Babesiosis. Human babesiosis is a tick-borne disease primarily associated with two protozoa of the family Piroplasmordia: *Babesia microti* and *Babesia divergens*, although other newly recognized species may also cause human infection. In 2018, there were 2,160 cases of human babesiosis reported in the United States.[4] The disease is a malaria-like syndrome characterized by fever, fatigue, and hemolytic anemia lasting from several days to a few months. In terms of clinical manifestations, babesiosis may vary widely, from asymptomatic infection to a severe, rapidly fatal disease.

Tularemia. Tularemia, sometimes called *rabbit fever* or *deer fly fever*, is a bacterial zoonosis that occurs throughout temperate climates of the northern hemisphere. Approximately 150 to 300 cases occur in the United States each year, but most cases occur in Arkansas, Missouri, and Oklahoma.[4,14] The disease may be contracted in a variety of ways: food, water, mud, articles of clothing, and (particularly) arthropod bites such as from ticks and certain biting flies.

Viruses associated with ticks. Tick-borne encephalitis (TBE) is a disease complex encompassing at least three syndromes caused by closely related viruses spanning from the British Isles (Louping ill), across Europe (Central European tick-borne encephalitis), to far-eastern Russia [Russian spring-summer encephalitis (RSSE)] (Figure 10.12). In Central Europe the typical case has a biphasic course with an early,

Figure 10.12 Approximate geographic distribution of tick-borne encephalitis.

viremic, flulike stage, followed about a week later by the appearance of signs of meningoencephalitis.[15] Central nervous system (CNS) disease is relatively mild, but occasional severe motor dysfunction and permanent disability occur. Powassan encephalitis (POW)—also in the TBE subgroup—is a relatively rare infection of humans that mostly occurs in the northeastern United States and adjacent regions of Canada. Characteristically, there is sudden onset of fever with temperature up to 40°C along with convulsions. Also, accompanying encephalitis is usually severe, characterized by vomiting, respiratory distress, and prolonged, sustained fever. Cases of POW are still relatively rare in North America, although its reported incidence is increasing.[4,16] There were 21 cases reported in 2018.[4] Colorado tick fever is a moderate, self-limiting febrile tick-borne illness occurring in the Rocky Mountain region of the United States and Canada. The primary vector is *Dermacentor andersoni*. Small mammals such as ground squirrels and the ticks themselves serve as reservoirs of the virus. Crimean-Congo hemorrhagic fever is a rather serious tick-borne illness occurring in many countries in central and eastern Europe, Russia, China, India, Pakistan, the Middle East, and parts of Africa. Transmission is mainly by *Hyalomma marginatum* and other closely related species. Rabbits, cattle, and goats are believed to be the reservoir hosts. Kyasanur forest disease, transmitted primarily by *Haemaphysalis spinigera* and related species, occurs in southern India. The disease is believed to be contracted by people working in/near the Kyasanur forest or cattle grazing at the forest edge. In the last decade, several new tick-borne viruses have been identified. Heartland virus (a *Phlebovirus*) is associated with the Lone Star tick, *Amblyomma americanum* and has been recognized in Missouri, Oklahoma, Kentucky, and Tennessee.[17,18] Only about 50 cases of Heartland virus have been identified. A couple of cases of a new *Thogotovirus* called Bourbon virus have been identified in the Midwest and southern United States with an unknown tick vector.[19] Evidence suggests the lone star tick may be a potential vector.[20]

Tick paralysis. Tick paralysis is characterized by an acute, ascending, flaccid motor paralysis that may terminate fatally if the tick is not located and removed. The causative agent is believed to be a salivary toxin produced by ticks when they feed. Many hard tick species may be involved, but *Dermacentor andersoni*, *D. variabilis*, and *Ixodes holocyclus* are notorious offenders. The disease is especially common in Australia. In North America, hundreds of cases have been documented from the Montana–British Columbia region.[21,22] Tick paralysis may occur in the southeastern United States as well. The author has seen two documented cases in young children admitted to the University of Mississippi Medical Center in Jackson, MS. Sporadic cases may occur elsewhere, such as Europe, Africa, and South America.

MAJOR TICK-BORNE DISEASES WORLDWIDE

- Lyme disease
- Rocky Mountain spotted fever, boutonneuse fever, and other related rickettsial diseases
- Ehrlichiosis (several types)
- Anaplasmosis
- Babesiosis
- Tick-borne relapsing fever
- Colorado tick fever
- Encephalitis viruses (tick-borne encephalitis and related diseases)
- Tularemia
- Tick paralysis

IPM and Alternative Control Methods

Integrated tick management (ITM) means using several different methods to reduce a tick population by weighing and balancing ecologic, economic, and epidemiologic costs and benefits of those methods. As is the case with classical IPM, one of the main goals is to minimize risk of ticks and tick-borne diseases to people and animals.[23] For example, careful landscaping may reduce tick and host animal habitats near the home (see next section), as well as controlled burning. The use of fire in forests has been shown to reduce many species of ectoparasites in the area, and *greatly* reduce the number of ticks, at least for a year or so.[24] Another component of ITM is management or treatment of host animals, both domestic and in wildlife by using the lowest effective pesticide dose, applying it only to animals at risk of infestation, and spraying habitats that are at the highest risk for tick invasion. One recent study comparing a traditional synthetic pyrethroid insecticide, a fungus-based acaricide, and a natural product-based acaricide against nymphal stage deer ticks and lone star ticks showed that the natural products failed to maintain 90% control through the study period.[25] In contrast, the traditional, residual insecticide suppressed 100% of deer ticks and >96% of lone star ticks, demonstrating effectiveness of chemical control of ticks.[25]

Due to clustering effects of hard ticks in nature, pesticides can be "precision targeted" toward tick hotspots and not broadcast over large areas. For example, in a study of lone star ticks (LST) in field plots in central Mississippi, the majority of nymphal stage ticks were collected in only 9.7% of the area,[26] indicating that spraying only 10% of the area could likely control a majority of ticks. A similar study with LSTs in northern Mississippi revealed spots of tick clustering along trails, along with long stretches of trail not having any ticks (Figure 10.13).[27]

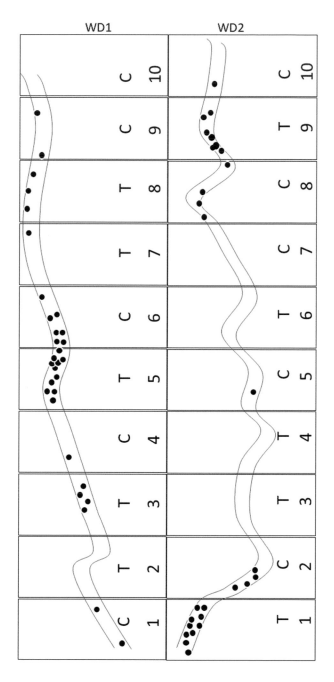

Figure 10.13 Trails in the woods showing spots of tick clustering. Each dot represents one or more tick collections at that spot.

Source: Figure copyright 2019 by Jerome Goddard, Ph.D.

Certainly any tick control strategy should be used in conjunction with personal protective measures such as boots and repellents. The ultimate goal is to reduce the number of cases of disease in people and animals with resources available. An integrated management approach does not preclude use of pesticides, but seeks to use chemicals effectively and responsibly in order to minimize exposure to people, pets, and the environment.

Landscape management. Research and computer models have shown that pesticides are the most effective way to reduce ticks, particularly when combined with landscaping changes that decrease tick habitat in frequently used areas of the yard. Also, most ticks (82%) are located within 10 meters of the lawn perimeter particularly along woodlands, stonewalls, or ornamental plantings.[28] Tick abundance in manicured lawns can be influenced by the amount of canopy vegetation and shade. Groundcover vegetation can harbor ticks. Woodland paths also may have a high number of ticks along the adjacent grass and bushes. The highest risk for ticks in suburban areas is the lawn perimeter in a zone formed by brushy areas, groundcover vegetation, and woods. The idea for residential tick management is to create a tick managed area around a home that incorporates portions of the yard that the family uses most frequently. This includes walkways, areas used for recreation, play, eating or entertainment, the mailbox, storage areas, and gardens.

Some actions to consider for ITM:

- Keep grass mowed.
- Remove leaf litter, brush and weeds at the edge of the lawn.
- Restrict use of groundcover in areas frequented by people and pets.
- Remove brush and leaves around fences and wood piles.
- Discourage rodent activity.
- Clean up and seal wood fences and small openings around the home.
- Move firewood piles and bird feeders away from the house.
- Manage pet activity; keep dogs and cats out of the woods to reduce ticks brought back into the home.
- Use plantings that do not attract deer or exclude deer through various types of fencing.
- Move children's swing sets and sand boxes away from the woodland edge and place them on a wood chip or mulch type foundation.
- Trim tree branches and shrubs around the lawn edge to let in more sunlight.
- Adopt landscaping techniques that use gravel pathways and mulches requiring less water application.
- Create a 3-foot or wider wood chip, mulch, or gravel border between lawn and woods.
- Consider using decking, tile, gravel, or container plantings in areas by the house or areas that are used frequently.

WHAT'S IN A NAME?

CHANGING THE NAME OF THE CAUSATIVE
AGENT OF LYME DISEASE

In 1981, Dr. Willy Burgdorfer discovered the causative agent of a newly recognized tick-borne disease occurring in the community of Old Lyme Connecticut. That disease agent was ultimately named after him, *Borrelia burgdorferi*. Drs. Mobalaji Adeolu, Radhey Gupta, and colleagues (now including one of the original authors of the Lyme disease agent paper, Dr. Alan Barbour at the University of California Irvine) have proposed a separation of the LD-causing spirochetes into a new genus, *Borreliella*,[1] based on two primary lines of evidence. First, phylogenetic data showing high evolutionary support for the new grouping, and second, consistent identification of conserved signature proteins (CSPs) and conserved signature insertions and deletions (INDELs) in each group, further supporting the genus separation. An obvious difference—as they would say—is that the two clades (groups) cause two different human infections with obviously different clinical and laboratory findings; however, there are other biological and phenotypic characteristics clearly separating the two groups. For example, *Borrelia* organisms are present at all times in the salivary glands of their tick vectors, while *Borreliella* only move to the salivary glands during the feeding process (they ordinarily reside in the midgut). Further, in *Borrelia* there is transovarial transmission (mother to offspring) of spirochetes, which does not occur in *Borreliella*. There has been considerable push-back on this proposition over the last 4 years, including a bewildering assortment of arguments with all kinds of molecular machinations. Laypeople, even clinicians, can get lost in the technical jargon and fail to see the importance (or not) of the arguments and their clinical relevance. As best this author can tell, there is ample evolutionary evidence to support the aforementioned new classification. Nonetheless, arguments can be made against changing the name due to public health and patient safety concerns that have been raised, including the following.[2]

1. The terms, Lyme disease and *Borrelia burgdorferi*, are intertwined in medical references worldwide.
2. Worldwide databases store diagnostic and treatment information. Changing the name might confound medical literature searches.

3. In some countries, payment for treatment and laboratory services may be stalled due to the name change.
4. Lyme disease is a professional risk in many countries for persons working in forested areas. Insurance companies may delay payment if diagnostic services are confounded by a name change.

The following is a response from Dr. Barbour himself on a "spirochete listserv" concerning the controversy: "*Borreliella burgdorferi* is a valid published name, just as *Borrelia burgdorferi* is a valid published name. *Borreliella* cannot be 'rescinded,' as some have insisted. One may choose (as I'm sure many will) not to use it and instead keep using the older name, *Borrelia burgdorferi*. That is their right, as is the right for someone to legitimately use *Borreliella burgdorferi* where that would be suitable. A journal or a reviewer for a journal cannot force someone not to use *Borreliella*, just as authors cannot be restricted from using *Borrelia burgdorferi*. There is no august official court somewhere that will lay down a final ruling on this (see the principles and rules governing naming of prokaryotic organisms), let alone some sort of straw poll on a listserv to settle it. There can be differences in taxonomic opinions between reasonable people. That's science. It will likely come down to usage of the name(s) over time." Taxonomic changes happen all the time in biology. Clinicians should be aware of this particular one since Lyme disease is such a sensational disease and patients or advocacy groups may bring it up. Nonetheless, the whole thing is probably a "non-issue" since the National Institutes of Health GenBank has apparently accepted *Borreliella*, as well as Bergey's Manual of Systematic Bacteriology.

[1] **Barbour AG, Adeolu M, Gupta RS**. Division of the genus *Borrelia* into two genera (corresponding to Lyme disease and relapsing fever groups) reflects their genetic and phenotypic distinctiveness and will lead to a better understanding of these two groups of microbes (Margos et al. (2016). There is inadequate evidence to support the division of the genus *Borrelia*. *Int J Syst Evol Microbiol*. doi: 10.1099/ijsem.0.001717).

[2] **Stevenson B, Fingerle V, Wormser GP, Margos G**. Public health and patient safety concerns merit retention of Lyme borreliosis-associated spirochetes within the genus *Borrelia*, and rejection of the genus novum *Borreliella*. *Ticks Tick Borne Dis*. 2019;10(1):1–4.

References

1. Barbour AG, Adeolu M, Gupta RS. Division of the genus *Borrelia* into two genera corresponding to Lyme disease and relapsing fever groups reflects their genetic and phenotypic distinctiveness and will lead to a better understanding of these two groups of microbes (rebuttal letter). *Int J Sys Evol Microbiol*. 2017;67:2058–2067.

2. Barbour AG, Gupta RS. The family Borreliaceae (Spirochaetales), a diverse group in two genera of tick-borne spirochetes of mammals, birds, and reptiles. *J Med Entomol.* 2021;doi: 10.1093/jmc/tjab055:1–12.

3. CDC. Lyme disease, United States, 2001–2002. In: CDC, MMWR. 2004;53:365–368.

4. CDC. Summary of notifiable diseases. In: National Notifiable Diseases Surveillance System (NNDSS); 2018. wwwn.cdc.gov/nnds/infections-tables-html.

5. Kugeler KJ, Schwartz AM, Delorey MJ, Mead PS, Hinckley AF. Estimating the frequency of Lyme disease diagnoses, United States, 2010–2018. *Emerg Inf Dis.* 2021;27:616–619.

6. Goddard J, Varela-Stokes A, Finley RW. Lyme-disease-like illnesses in the South. *J Mississippi State Med Assoc.* 2012;53(3):68–72.

7. Goddard J. Not all erythema migrans lesions are Lyme disease. *Am J Med.* 2016;epub ahead of print, doi: 10.1016/j.amjmed:doi: 10.1016/j.amjmed.

8. Shapiro ED. Lyme disease in 2018: what is new (and what is not). *JAMA.* 2018;Electronic publication ahead of print, August 2, 2018.

9. CDC. Caution regarding testing for Lyme disease. In: CDC, MMWR. 2005;54:125.

10. FDA. Beware of ticks and Lyme disease. In: U.S. Food and Drug Administration Consumer Health Information, update on Lyme disease; 2007. www.fda.gov/consumer/updates/lymedisease062707.html.

11. Wilske B, Fingerle V, Schulte-Spechtel U. Microbiological and serological diagnosis of Lyme borreliosis. *FEMS Immunol Med Microbiol.* 2007;49(1):13–21.

12. Dumler JS, Bakken JS. Ehrlichial diseases of humans: emerging tick-borne infections. *Clin Infect Dis.* 1995;20:1102–1110.

13. Spach DH, Liles WC, Campbell GL, Quick RE, Anderson DEJ, Fritsche TR. Tick-borne diseases in the United States. *N Engl J Med.* 1993;329:936–947.

14. Monath TP, Johnson KM. Diseases transmitted primarily by arthropod vectors. In: Last JM, Wallace RB, Eds., eds. *Public Health and Preventive Medicine.* 13th ed. Norwalk, CT: Appleton and Lange; 1992.

15. Nuttall PA, Labuda M. Tick-borne encephalitis subgroup. In: Sonenshine DE, Mather TN, eds. *Ecological Dynamics of Tick-borne Zoonoses.* New York: Oxford University Press; 1994:351.

16. McMullan LK, Folk SM, Kelly AJ, et al. A new Phlebovirus associated with severe febrile illness in Missouri *N Engl J Med.* 2012;367(9):834–841.

17. Pastula DM, Turabelidze G, Yates KF, et al. Heartland virus disease—United States, 2012–2013. In: CDC, MMWR. 2014;63:270–271.

18. Kosoy OI, Lambert AJ, Hawkinson DJ, et al. Novel thogotovirus associated with febrile illness and death, United States, 2014. *Emerg Infect Dis.* 2015;21(5):760–764.

19. Savage HM, Burkhalter KL, Godsey MSJ, et al. Bourbon virus in field-collected ticks, Missouri, USA. *Emerg Infect Dis.* 2017;23(12):2017–2022.

20. Gregson JD. Tick paralysis: an appraisal of natural and experimental data. In: Canada Dept. Agri. Monograph No. 9; 1973:48.

21. Schmitt N, Bowmer EJ, Gregson JD. Tick paralysis in British Columbia. *Canadian Med Assoc J.* 1969;100:417–421.

22. USDA. A national roadmap for integrated pest management. In: The Federal IPM Coordinating Committee Report; 2018. www.ars.usda.gov/ARSUserFiles/OPMP/IPM%20Road%20Map%20FINAL.pdf.

23. Scasta JD. Fire and parasites: an under-recognized form of anthropogenic land use change and mecahnism of disease exposure. *EcoHealth.* 2015; doi: 10.1007/s10393-015-1024-5.

24. Schulze TL, Jordan RA. Synthetic pyrethroid, natural product, and entomo-pathogenic fungi acaricide product formulations for sustained early season suppression of host-seeking *Ixodes scapularis* and *Amblyomma americanum* nymphs. *J Med Entomol.* 2020;Epub ahead of print, doi: 10.1093/jme/tjaa248.
25. Goddard J. Clustering effects of lone star ticks in nature: implications for control. *J Environ Health.* 1997;59:8–11.
26. Goddard J. Tick control in state parks. In: Robinson, W.H., Rettich, F., and Rambo, G.W. eds. Proceed. 3rd Internat. Conf. Urban Pests, Prague, Czech Republic; 1998:485–492.
27. Stafford KC. Tick management handbook. In: Connecticut Agri. Exp. Sta. Bull. No. 1010; 2007:78 pp.

chapter eleven

Fleas

Importance and Physical Description

Fleas are small, laterally flattened, wingless insects that are of great importance as vectors of disease in many parts of the world. They have complete metamorphosis, with egg, larval, pupal, and adult stages (Figure 11.1). Adult fleas are between 2 and 6 mm long, and are usually brown or reddish brown with stout spines on their head and thorax (Figure 11.2).[1] They have a short, clublike antenna over each eye. Each segment of their three-segmented thorax bears a pair of powerful legs terminating in two curved claws. Most fleas can move quickly on skin or in hair and may jump 30 cm or more. They are readily recognized by their jumping behavior when disturbed.

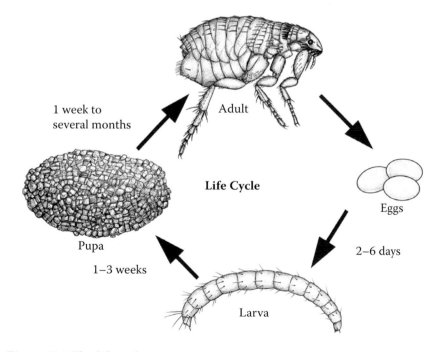

1 week to several months

Adult

Life Cycle

Eggs

Pupa

2–6 days

1–3 weeks

Larva

Figure 11.1 Flea life cycle.

Source: Courtesy of the Mississippi State University Extension Service and Joe MacGown

Figure 11.2 Typical adult flea.

Biology and Life Cycle

The life cycle of the cat flea is presented here as an example (Figure 11.1). Adult female fleas begin laying eggs 1–4 days after blood feeding. Bloodmeals are commonly obtained from medium-sized mammals, such as cats, dogs, raccoons, and opossums, but humans may be utilized as well. Females lay 10–20 eggs a day and may produce several hundred eggs in their lifetime. Eggs are normally deposited in nest litter, bedding, carpets, or other resting sites. Warm, moist conditions are needed for egg production. After the egg stage, spiny, yellowish-white larvae emerge which have chewing mouthparts and feed on host-associated debris, including food particles, dead skin, and feathers. Interestingly, blood spots defecated by adult fleas also serve as a food source for larvae. There are three larval molts (instars) prior to pupating. Flea larvae are prone to desiccation and will quickly die if continuously exposed to <60–70% relative humidity. For the pupal stage, flea larvae spin a loose silken cocoon interwoven with debris. During unfavorable environmental conditions, or if hosts are not available, developing adult fleas may remain inactive within the cocoon for extended periods. Subsequently, adult emergence from the cocoon may be triggered by vibrations resulting from host animal movements.

Distribution

Fleas occur as ectoparasites on a wide variety of birds and mammals worldwide.

Impact on Human Health

Fleas bite people and wild and domestic animals causing irritation, blood loss, and severe discomfort. They may serve as intermediate hosts for certain helminths (Figure 11.3), as well as transmit various disease agents

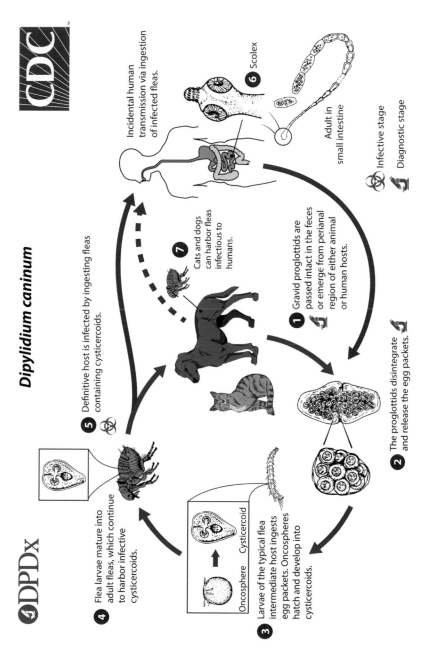

Figure 11.3 Fleas may serve as intermediate hosts of the dog tapeworm, *Diphylidium caninum*.

Source: From the Centers for Disease Control

(see the following). The flea-bite lesion begins as a punctate hemorrhagic area representing the site of probing by the insect which may have a center elevated into a papule, vesicle, or even a bulla. Lesions may occur in clusters as the flea explores the skin surface, frequently stopping and probing. A wheal usually forms around each probe site with the wheal reaching its peak in 5 to 30 minutes. Itching is almost always present. Generally, there is transition to an indurated papular lesion within 12 to 24 hours. Immunologically, flea bites may produce both immediate and delayed skin reactions.[2,3] In sensitized individuals, the delayed reaction appears in 12 to 24 hours and can persist for a week or more. This delayed papular reaction with intense itching is most often the reason people present to clinics.

Plague. Plague, sometimes called black death, a zoonotic disease caused by the bacterium *Yersinia pestis*, has been associated with humans since recorded history, causing devastating effects on human civilization. For example, in the 14th century approximately 25 million people died of plague in Europe.[4] Even though treatable with antibiotics, to this day, there are still hundreds of cases occurring annually over much of the world (Figure 11.4). In the United States, from 1970 to 1994 a total of 334 cases of indigenous plague were reported; the peak years were 1983 and 1984, in which there were 40 and 31 cases, respectively.[5] In 2015, 16 cases of plague were reported to the CDC from Arizona, California, Colorado, Georgia, Michigan, New Mexico, Oregon, and Utah.[6] Sylvatic plague, sometimes also called campestral plague, is ever present in endemic areas, circulating among rock and

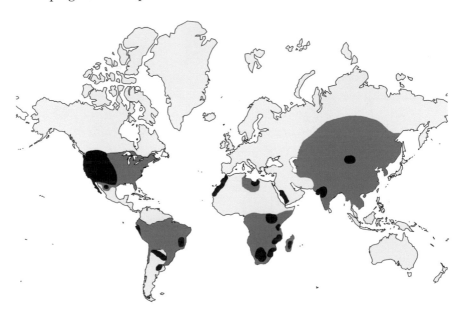

Figure 11.4 Approximate geographic distribution of plague. Tan = Countries Reporting Plague; Dark Red = Foci of Sylvatic Plague.

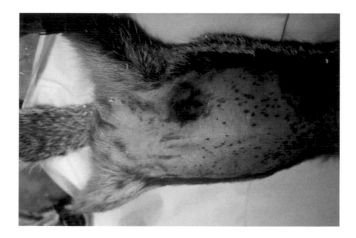

Figure 11.5 Shaved underside of a rock squirrel infected with *Yersinia pestis*, showing petechial rash similar in appearance to that found on infected humans.

Source: From the Centers for Disease Control, photo by William Archibald

ground squirrels, deer mice, voles, chipmunks, and others (Figure 11.5). Transmission from wild rodents to humans can occur by direct contact with sick or dead animals, but is rare. For example, in 2012 a 7-year-old girl contracted plague after playing with a dead ground squirrel in Colorado.[7]

Three clinical forms of plague are recognized: bubonic, septicemic, and pneumonic. The septicemic and pneumonic forms are usually secondary to the bubonic form, and the bubonic form is the most common in the Americas. Pneumonic plague is the most dangerous because of its rapid spread by aerosols (coughing).

Murine typhus. The rickettsial disease called murine typhus is transmitted to humans by fleas and is characterized by headache, chills, prostration, fever, and general pains. There may be a macular rash, especially on the trunk. Infection occurs when infectious flea feces are rubbed into the fleabite wound or other breaks in the skin. Murine typhus is usually mild with negligible mortality, except in the elderly, although severe cases occasionally occur with hepatic and renal dysfunction. For many years, there were only about 100 cases of murine typhus reported annually in the United States.[8,9] However, a recent paper reported 90 cases of murine typhus from just 2 Texas hospitals over a 3-year period.[10] The classic cycle involves rat-to-rat transmission with the oriental rat flea, *Xenopsylla cheopis*, being the main vector; however, the classic cycle seems to have been replaced in suburban areas by a peridomestic animal cycle involving free-ranging cats, dogs, opossums, and their fleas.[11] Murine typhus is one of the most widely distributed arthropod-borne infections endemic in many coastal areas and ports throughout the world.[11]

Prevention and Control

Fleas infesting a house or property should first be correctly identified as this will inform pest control personnel of their biology, host(s), and habitats. Preventive measures for fleas consist of denying wild and domesticated animals access into the structure. This includes blocking access of wild animals into the crawl space (if there is one). Sometimes flea problems are the result of wild animals present on a property or inside a structure, even if there are no pet animals present. If pets are present, they need to be treated for fleas, preferably by a veterinarian or with on-animal products prescribed by a veterinarian. Indoor control of fleas is generally accomplished by spraying an appropriately labeled pesticide and/or an insect growth regulator (IGR) either as a spot treatment or a wide area treatment (Note: not all products are labeled for wide area application). The key to successful flea control is to be thorough in cleaning and treating all areas in the structure where fleas occur, giving special attention to areas where pets sleep or rest.[12] Outdoor treatment for fleas is by application of residual liquid or granular pesticides using power equipment.

IPM and Alternative Control Methods

IPM for fleas includes plugging or sealing holes providing access into the structure using cement, caulk, or quick-drying foam. Ensuring doors and windows are tight-fitting and screened can also help prevent flea access. Indoor IPM methods include vacuuming all rugs, floors, and fabric-covered furniture. Infested pet bedding must be either cleaned, vacuumed, or disposed of. Outdoor flea IPM includes keeping the grass and weeds cut and elimination of alternate hosts and their harborages within 150 meters of the structure.

References

1. Lewis RE. Resume of the Siphonaptera. *J Med Entomol*. 1998;35:377–389.
2. Alexander JO. *Arthropods and Human Skin*. Berlin: Springer-Verlag; 1984.
3. Feingold BF, Benjamini E. Allergy to flea bites. *Ann Allergy*. 1961;19:1275–1279.
4. Mullett CF. *The Bubonic Plague and England*. Lexington, KY: University of Kentucky Press; 1956.
5. Craven RB, Maupin GO, Beard ML, Quan TJ, Barnes AM. Reported cases of human plague infections in the U.S. *J Med Entomol*. 1993;30:758–761.
6. CDC. Summary of notifiable infectious diseases and conditions—United States, 2015. In: CDC, MMWR. 2017;64(53):1–144.
7. Golgowski N. Joy for girl, 7, who caught bubonic plague from dead squirrel. In: The Daily Mail (UK), September 9 issue; 2012. www.dailymail.co.uk/news/article-2200890/Sierra-Jane-Downing-Girl-7-caught-bubonic-plague-dead-squirrel-camping-trip-leave-hospital.html.

8. Civen R, Ngo V. Murine typhus: an unrecognized suburban vectorborne disease. *Clin Infect Dis*. 2008;46:913–918.

9. Blanton LS, Vohra RF, Bouyer D, Walker DH. Reemergence of murine typhus in Galvaston, Texas, USA, 2013. *Emerg Infect Dis*. 2015;21:484–486.

10. Afzal Z, Kallumadanda S, Wang F, Hemmige V, Musher D. Acute febrile illness and complications due to murine typhus, Texas, USA. *Emerg Infect Dis*. 2017;23:1268–1273.

11. Azad AF, Radulovic S, Higgins JA, Noden BH, Troyer JM. Flea-borne rickettsioses: ecologic considerations. *Emerg Infect Dis*. 1997;3:319–327.

12. Layton B, Goddard J, Edwards KT, MacGown JA. Control fleas on your pet, in your house, and in your yard. In: Mississippi State University Extension Service, Publication Number 2597; 2019:12 pp.

chapter twelve

Lice

Importance and Physical Description

Bloodsucking lice are parasites of humans and many other mammals and birds worldwide, and certain species are significant vectors of disease agents. Human body lice and head lice are almost identical in appearance; however, body lice are usually about 15–20% larger than head lice. They are tiny (2–4 mm long), elongate, soft-bodied, light-colored, wingless insects that are dorsoventrally flattened, with an angular ovoid head and a nine-segmented abdomen (Figure 12.1). The eggs are small (about 1 mm), oval, white or cream-colored objects with a distinct cap on one end. Eggs are attached to clothing in the case of body lice and to hairs in the case of head and pubic lice (Figure 12.2). The head bears a pair of simple lateral eyes and a pair of short five-segmented antennae. Pubic lice are dark gray to brown in color and are called *crab lice* because of their crablike shape (Figure 12.3). They are distinctly flattened, oval, and much wider than body or head lice. As with head and body lice, the head bears a pair of simple lateral eyes and a pair of short five-segmented antennae. They are 1.5–2.0 mm long; their second and third legs are enlarged and contain a modified claw with a thumblike projection, which aids them in grasping hair shafts.

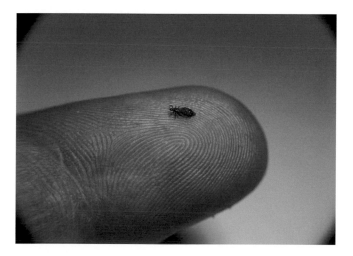

Figure 12.1 Body louse on fingertip.

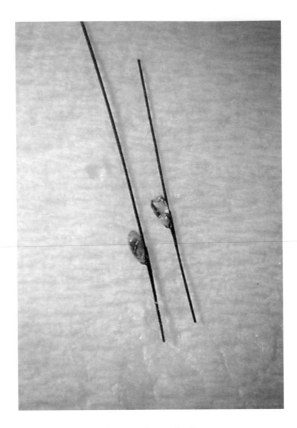

Figure 12.2 Head lice eggs attached to hair shafts.

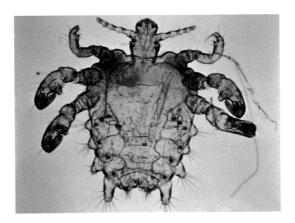

Figure 12.3 Public louse adult.

Source: From the Centers for Disease Control

Distribution

Human body, head, and pubic lice occur on humans worldwide.

Impact on Human Health

Head lice are not known to transmit disease agents, but their bites may cause considerable nuisance and irritation. Body lice are known to transmit the agent of *epidemic typhus, Rickettsia prowazeki,* and there have been devastating epidemics of this disease in the past. Typhus is still endemic in poorly developed countries where people live in filthy, crowded conditions (Figure 12.4). Besides louse-borne typhus, body lice transmit the agents of trench fever and epidemic relapsing fever. *Trench fever,* caused by *Bartonella quintana,* was a significant source of morbidity and mortality during WWI[1] (Figure 12.5), and is still widespread in parts of Europe, Asia, Africa, Mexico, and Central and South America, mainly in an asymptomatic form. However, it is now recognized as a reemerging pathogen among homeless populations in the United States and Europe, where it is responsible for a wide spectrum of conditions such as chronic bacteremia, endocarditis, and bacillary angiomatosis.[2-4] *Louse-borne relapsing fever* (LBRF) occurs primarily in the Horn of Africa, but may be seen in travelers, refugees, and immigrants from that area.[5] There were millions of cases of LBRF during the two World Wars of the 20th century,[6] and

Figure 12.4 Approximate geographic distribution of epidemic typhus.

Figure 12.5 Soldiers in WWI were affected by trench fever.

Source: From the National Library of Medicine

4,972 cases with 29 deaths were reported worldwide in 1971.[7] Pubic lice are not known to transmit disease organisms. However, the condition, pediculosis pubis, frequently coexists with other venereal diseases, particularly gonorrhea and trichomonas. One study indicated that one-third of patients with pubic lice may have other sexually transmitted diseases.[8]

Prevention and Control

Management of head lice infestations requires three general steps: (1) delousing infested individuals, with retreatment as necessary; (2) removing nits from the hair as thoroughly as possible; and (3) delousing personal items (clothes, hats, combs, pillows, etc.). An important principle of head lice management is to treat all infested members of a family concurrently. If an infested school-age child is the only family member treated, he or she may be quickly reinfested by a sibling or parent who is also unknowingly infested. Individuals with head lice should be treated with one of the approved pediculicidal shampoos or cream rinses. Finally, efforts should be made to delouse personal belongings of infested individuals. Washable clothing, hats, bedding, and other personal items should be washed properly and dried in a clothes dryer for at least 20 to 30 minutes.

Because body lice infest both the patient and his or her clothing, control strategies involve frequently changing clothing, washing infested garments in very hot water or having them dry-cleaned, and using pediculicidal lotions or shampoos. The primary element of control is to ensure that all clothing and bedding of infested persons is sanitized or treated. Clothing and bedding can also be disinfected by spraying with pyrethrin preparations or other approved insecticides or dusts. When mass treatments are indicated (such as in a war or natural disaster), insecticide dusts may be applied directly to the body. Newer treatment strategies may involve use of oral ivermectin.

As pubic lice infestations are usually transmitted through sexual contact, it is important to have the sexual contacts of the infested person examined and treated if needed. Likewise, as some family members all sleep in the same bed, if one member of a family has an infestation then all family members should be examined and infested ones treated. As with head and body lice control products, some are sold over-the-counter and some are by prescription only.

References

1. Durden LA, Hinkle NC. Fleas (Siphonaptera). In: Mullen GR, Durden LA, eds. *Medical and Veterinary Entomology*. 2nd ed. New York: Elsevier; 2009:115–135.
2. Foucault C, Brouqui P, Raoult D. *Bartonella quintana* characteristics and clinical management. *Emerg Infect Dis*. 2006;12:217–223.
3. Voelker R. Lice-borne diseases in the homeless population. *JAMA*. 2014;312:1962.
4. Bonilla DL, Cole-Porse C, Kjemtrup A, Osikowicz L, Kosoy M. Risk factors for human lice and bartonellosis among the homeless, San Francisco, California, USA. *Emerg Infect Dis*. 2014;20(10):1645–1651.
5. von Both U, Alberer M. *Borrelia recurrentis* infection. *N Eng J Med*. 2016;375:e5.
6. Barbour A. Relapsing fever and other *Borrelia* diseases. In: Guerrant RL, Walker DH, Weller PF, eds. *Tropical Infectious Diseases: Principles, Pathogens, and Practice*. 3rd ed. New York: Saunders Elsevier; 2011:295–302.
7. Harwood RF, James MT. *Entomology in Human and Animal Health*. 7th ed. New York: Macmillan; 1979.
8. Chapel TA, Katta T, Kuszmar T. *Pthrirus pubis* in clinic for treatment of sexually transmitted diseases. *Sex Trans Dis*. 1979;6:257–261.

chapter thirteen

Sand Flies

Importance and Physical Description

Sand flies are small and delicate bloodsucking flies (Figure 13.1) in the family Psychodidae that transmit the causative agents of bartonellosis (Carrión's disease), sand fly fever, and leishmaniasis. Aside from their disease transmission potential, sand flies may cause considerable irritation by their biting alone (nuisance effect). Adult flies are long-legged, about 3 mm long, and golden, brownish, or gray in color. Females have long, piercing mouthparts adapted for bloodsucking. One way to distinguish sand flies from other similarly shaped gnats is the way they hold their wings V-shaped at rest. They have long, multisegmented antennae and hairs (not scales) covering much of the body and wing margins.

Figure 13.1 Adult sand fly.

Source: From the Centers for Disease Control, photo by Dr. Hadley Collins

Distribution

Sand flies occur in tropical, subtropical, and temperate areas worldwide. Generally, members of the sand fly genera *Phlebotomus* and *Sergentomyia* occur in the Old World, while *Lutzomyia*, *Brumptomyia*, and *Warileya* occur in the tropics and subtropics of the New World. *Phlebotomus papatasi* is one of the most important sand fly species and is a widespread vector of zoonotic cutaneous leishmaniasis in the Old World. *Lutzomyia longipalpis* is a major vector of leishmaniasis in the New World.

Impact on Human Health

Bartonellosis is a disease caused by the bacterium, *Bartonella bacilliformis*, which occurs in the mountain valleys of Peru, Ecuador, and southwest Colombia. *Sand fly fever*, a viral disease, occurs in those parts of southern Europe, the Mediterranean, the Near and Middle East, Asia, and Central and South America where their *Phlebotomus* vectors exist. *Leishmaniasis* is a hugely significant parasitic disease occurring in tropical and subtropical areas over much of the world, causing much morbidity and mortality annually. There has been a general resurgence of leishmaniasis worldwide. The disease is now found in at least 88 countries (Figures 13.2–13.3) and is increasingly being reported in nonendemic areas. Global incidence

Figure 13.2 Approximate geographic distribution of cutaneous leishmaniasis.

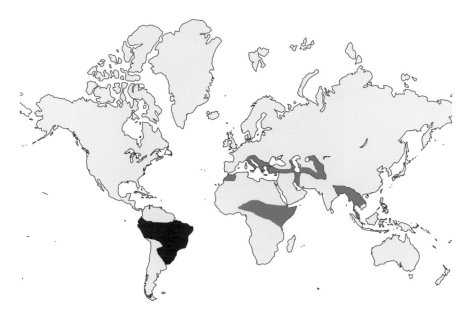

Figure 13.3 Approximate geographic distribution of visceral and mucocutaneous leishmaniasis. Tan = Visceral; Dark Red = Muco-cutaneous and Visceral Forms.

is estimated to be 700,000–1,000,000 new cases reported annually with more than 20,000 deaths.[1] Clinically, leishmaniasis manifests itself in four main forms: (1) cutaneous, (2) mucocutaneous, (3) diffuse cutaneous, and (4) visceral. The cutaneous form may appear as small to moderately sized and self-limiting ulcers that are slow to heal (Figure 13.4–13.5). When there is nasal and oral mucosal involvement resulting in tissue destruction, the disease is labeled mucocutaneous leishmaniasis. Sometimes there are widespread cutaneous papules or nodules all over the body, a condition termed diffuse cutaneous leishmaniasis. Last, when the parasites invade cells of the spleen, bone marrow, and liver, the disease is called visceral leishmaniasis. There have been reports of visceral leishmaniasis in foxhounds in the United States,[2] but the only human sand fly-transmitted disease in the United States at this time is cutaneous leishmaniasis (a few cases) diagnosed each year in south Texas and Arizona.[3–6]

Prevention and Control

Although the most serious problems from sand flies occur in tropical countries, there are many species in temperate zones, as well. The author often collects them in mosquito light traps in Mississippi. Because sand flies do not bite through clothing, long sleeves, trousers, and socks should

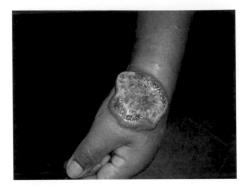

Figure 13.4 Cutaneous leishmaniasis.

Source: Courtesy of the U.S. Armed Forces Pest Management Board, Combat Aviation Advisor, 6th Special Operations Squadron, Hulburt Field, Florida

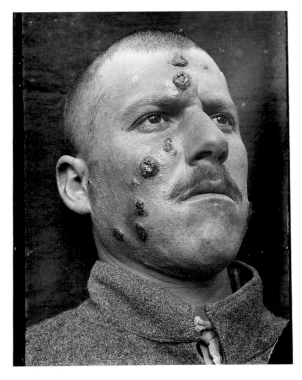

Figure 13.5 "Jericho buttons" (cutaneous leishmaniasis lesions), known for frequency of cases near the ancient city of Jericho.

Source: Photo courtesy the Matson Photograph Collection, U.S. Library of Congress, public domain

be worn in areas where sand flies are active. In heavily infested areas, head nets, gloves, and repellent-treated net jackets and hoods can provide additional protection. Campsites should be chosen that are high, breezy, open, and dry. Fine-mesh bed nets should be used in areas where sand flies are present.

References

1. WHO. Leishmaniasis. In: World Health Organization, Global Health Observatory Data; 2017. www.who.int/gho/neglected_diseases/leishmaniasis/en/.
2. Duprey ZH, Steurer FJ, Rooney JA, et al. Canine visceral leishmaniasis, United States and Canada, 2000–2003. *Emerg Infect Dis*. 2006;12:440–446.
3. Furner BB. Cutaneous leishmaniasis in Texas: report of a case and review of the literature. *J Am Acad Dermatol*. 1990;23:368–371.
4. Grimaldi G, Jr., Tesh RB, McMahon-Pratt G. A review of the geographic distribution and epidemiology of leishmaniasis in the New World. *Am J Trop Med Hyg*. 1989;41:687–693.
5. McHugh CP, Grogl M, Kreutzer RD. Isolation of *Leishmania mexicana* from *Lutzomyia anthophora* collected in Texas. *J Med Entomol*. 1993;30:631–633.
6. de Almeida ME, Zheng Y, Nascimento FS, et al. Cutaneous leishmaniasis caused by an unknown *Leishmania* strain, Arizona, USA. *Emerg Infect Dis*. 2021;27:1714–1960.

chapter fourteen

Tsetse Flies

Importance and Physical Description

Flies in the genus *Glossina* are called tsetse flies, which are vectors of several trypanosomes (protozoans) of people and animals. The main disease associated with tsetse flies is human African trypanosomiasis (HAT), caused by subspecies of the protozoan *Trypanosoma brucei*. A related disease of cattle is called Nagana. Other than the possibility for sleeping sickness transmission, bites by tsetse flies are generally only of minor consequence. However, some individuals may become sensitized to the saliva, leading to welts.

Tsetse flies look similar to honey bees and are 7–13 mm long and yellow, brown, or black in color. They have a long, slender proboscis that is held out in front of the fly at rest. In addition, the discal cell of the wing (first M) is shaped like a meat cleaver or hatchet (Figure 14.1). Tsetse flies fold their wings scissorlike over their back at rest, and the arista arising from the short, three-jointed antennae has rays that are branched bilaterally.

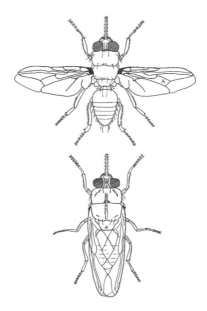

Figure 14.1 Adults tsetse fly; "hatchet cell" marked with an A.

Source: From the U.S. Department of Agriculture

233

Distribution

Tsetse flies are mostly found in tropical Africa between 15°N and 20°S latitude.

Impact on Human Health

HAT, or sleeping sickness, has a complex life cycle (Figure 14.2) and is characterized by a steady progression of meningoencephalitis, with increase of apathy, fatigability, confusion, and somnolence. The patient may gradually become more and more difficult to arouse and finally enters a coma. Cases of HAT continue to decrease due to intensive tsetse fly trapping and control. In 2009, the number of cases fell below 10,000 for the first time in 50 years,[1] then down to only about 3,000 in 2015.[2] Countries affected the most include the Democratic Republic of Congo, Angola, and Sudan. The vector biology and life cycle of HAT are quite complicated (see the overviews published previously).[3–5] Briefly, the chief vectors of *Trypanosoma brucei gambiense*, the cause of the Gambian form of sleeping sickness, are

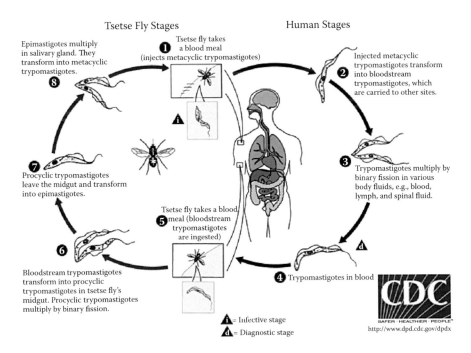

Figure 14.2 Life cycle of human African trypanosomiasis.

Source: From the Centers for Disease Control

Glossina palpalis, *G. fuscipes*, and *G. tachinoides*. Cases of Gambian sleeping sickness occur in Western and Central Africa and are usually more chronic (Figure 14.3). In Eastern and Southern Africa, the Rhodesian (or eastern) form, which is virulent and rapidly progressive, is caused by *T. brucei rhodesiense*. Uganda is the only country in which both subspecies are present. The primary vectors of the Rhodesian form are *G. morsitans*, *G. swynnertoni*, and *G. pallidipes*. The eastern form may be contracted by travelers on game safaris or eco-vacations.

Prevention and Control

There have been intensive efforts to control tsetse flies with insecticides and traps, which have yielded measurable successes. Since there are no vaccines or prophylactic drugs for African trypanosomiasis, the CDC recommends the following personal protective measures:

- Wear long-sleeved shirts and pants of medium-weight material in neutral colors that blend with the background environment. Tsetse flies are attracted to bright or dark colors, and they can bite through lightweight clothing.
- Inspect vehicles before entering. The flies are attracted to the motion and dust from moving vehicles.

Figure 14.3 Approximate geographic distribution of the two forms of African trypanosomiasis. Green = Gambian; Orange = Rhodesian; Dark Red = Both.

- Avoid bushes. The tsetse fly is less active during the hottest part of the day but will bite if disturbed.
- Use insect repellent. Permethrin-impregnated clothing and insect repellent have not been proved to be particularly effective against tsetse flies, but they will prevent other insect bites that can cause illness.

References

1. WHO. African trypanosomiasis. In: World Health Organization, Media Center, Fact Sheet Number 259; 2010:6 pp.
2. Buscher P, Cecchi G, Jamonneau V, Priotto G. Human African trypanosomiasis. *Lancet*. 2017;390(10110):2397–2409.
3. Nash TAM. A review of the African trypanosomiasis problem. *Trop Dis Bull*. 1960;57:973–980.
4. Pepin J, Donelson JE. African trypanosomiasis. In: Guerrant RL, Walker DH, Weller PF, eds. *Tropical Infectious Diseases*. 3rd ed. Philadelphia: Elsevier Saunders; 2011:682–688.
5. Willet KC. African trypanosomiasis. *Ann Rev Entomol*. 1963;8:197–213.

chapter fifteen

Black Flies

Importance and Physical Description

Black flies (also called *buffalo gnats*, *turkey gnats*, and *Kolumbtz flies*) are small, humpbacked flies that are severe nuisance pests and occasionally vectors of disease.[1,2] The primary disease associated with these pests is onchocerciasis in Africa and South America, caused by a nematode worm. In temperate regions, black flies are notorious pests often occurring in tremendous swarms and biting humans and animals viciously. There are reports of human deaths from black fly biting in the older literature, and one report of 400 mules dying within a few days after exposure to black fly swarms in Louisiana.[3,4] As opposed to mosquitoes (which breed in stagnant water), black flies breed in fast-flowing streams and rivers (Figure 15.1).

Black flies are smaller than mosquitoes, ranging in size from about 2–5 mm, with broad wings, stout bodies, and large compound eyes (Figure 15.2). Their antennae are short (although in 9–12 segments) and bare.

Figure 15.1 Black flies breed in fast-moving water.

Source: Photo copyright 2009 by Jerome Goddard, Ph.D.

Figure 15.2 Adult black fly.

Source: Courtesy of Jerome Goddard, Ph.D., and the Mississippi State University Extension Service

Distribution

There are numerous important species of black flies: *Prosimuliim mixtum* is a serious pest of people and animals in much of the United States as well as *Cnephia pecuarum* in the Mississippi Valley, and *Simulium meridionale* in the eastern and south-central United States. In Mississippi, recent outbreaks of *S. meridionale* have made life miserable for thousands of residents in the Mississippi Delta and killed many chickens, purple martins, and other bird species in the area.[5,6] *Simuliim vittatum* and *S. venustum* may seriously annoy livestock, fishermen, and campers in the northern United States. In south-central Europe there have been severe outbreaks of *S. colombaschense* (the infamous golubatz fly) and *S. erythrocephalum*. Other notorious pests in Europe include *S. equinum*, *S. ornatum*, and *S. reptans*. In Africa, members of the *S. damnosum* and *S. neavei* complexes are important vectors of onchocerciasis. In Central and South America, vectors of onchocerciasis are *S. ochraceum* and *S. metallicum*.

Impact on Human Health

In the tropics, black flies are vectors of the parasite *Onchocerca volvulus*, which causes a chronic nonfatal disease with fibrous nodules in subcutaneous tissues (*onchocerciasis*) and sometimes visual disturbances and blindness (*river blindness*) (Figure 15.3). The World Health Organization estimates that about 17 million people have onchocerciasis in Africa and Latin America.[7]

In other areas of the world, the main problem from black flies is nuisance biting, and these effects can be quite severe (Figure 15.4), limiting almost all outdoor activity during outbreaks. Bites may be painless at first, but bleed profusely due to salivary proteins that prevent clotting.[8]

Figure 15.3 River blindness caused by the filarial worm *Onchocerca volvulus* and transmitted by black flies.

Source: From the World Health Organization

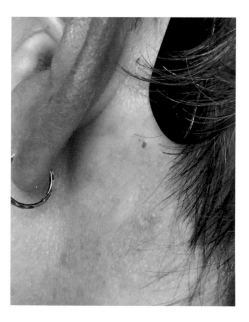

Figure 15.4 Black fly bite under ear.

Source: Photo courtesy Dr. Wendy C. Varnado, with permission

Systemic reactions to black fly bites have been reported, consisting of itching, burning, and papular lesions accompanied by fever, leukocytosis, and lymphadenitis. Satellite bubos have also been reported.[3] Death may result from anaphylactic shock, suffocation, and toxemia related to black fly bites.

Prevention and Control

Fortunately, black flies are daytime biters and rarely venture indoors; therefore, people can limit outdoor activity during peak black fly emergences. Providing shelters for backyard poultry may help protect them, since the flies do not like to enter enclosures. Repellents containing DEET have been reported effective for humans (but may need to be reapplied frequently), and wearing light-colored clothing may help keep the gnats away. Chemical control of black flies involves application of insecticides for both adults and larvae. This has only limited success, since it is often difficult to locate and treat all breeding sites. Larviciding with the "biological" control agent, *Bacillus thuringiensis israeliensis*, or BTI, (a spore-forming bacteria that kills the feeding larvae) has shown success in parts of the United States[9] and many African countries participating in the Onchocerciasis Control Program (OCP).

References

1. Crosskey RW. Blackflies. In: Lane RP, Crosskey RW, eds. *Medical Insects and Arachnids*. London: Chapman and Hall; 1996:241–287.
2. Nelson GS. Onchocerciasis. *Adv Parasitol*. 1970;8:173–177.
3. Stokes JH. A clinical, pathological, and experimental study of the lesions produced by the bite of the black fly, *Simulium venustum. J Cut Dis Incl Syph*. 1914;32(11):1–46.
4. Nations TM, Edwards KT, Goddard J. The George H. Bradley black fly papers. *Midsouth Entomol*. 2016;8:73–75.
5. Zema N. Buffalo gnats a temporary nuisance. In: The Natchez Democrat, Natchez, MS. Wednesday, June 1 issue; 2011.
6. Jones KH, Johnson N, Yang S, et al. Investigations into outbreaks of black fly attacks and subsequent avian haemosporidians in backyard type poultry and other exposed avian species. *Avian Dis*. 2014.
7. WHO. Progress report on the elimination of human onchocerciasis, 2016–2017. *Weekly Epidemiol Rec*. 2017;92:681–694.
8. Cupp EW, Cupp MS. Black fly salivary secretions: importance in vector competence and disease. *J Med Entomol*. 1997;34:87–94.
9. Clements R. Colton, NY "Black Fly Diva" battles the biting bugs for 35 years. In: St. Lawrence University, North Country Public Radio, Canton, New York; 2021. www.northcountrypublicradio.org/news/story/43888/20210618/colton-s-black-fly-diva-battles-the-biting-bugs-for-35-years.

chapter sixteen

Bed Bugs

Importance and Physical Description

The common bed bug, *Cimex lectularius*, has been a parasite of humans for thousands of years. Historically, the bloodsucking parasites were fairly common in human habitations worldwide where the little saying "sleep tight, don't let the bed bugs bite" meant something. Bed bugs had nearly disappeared in developing countries until recently, where, in the last three decades or so, they have been making a progressively rapid comeback. In many areas, they are now the number one urban pest.[1-4] Bed bugs have been suspected in the transmission of more than 40 disease organisms.[5] However, at this time there is little evidence bed bugs biologically transmit human pathogens,[6] with the possible exception of *Trypanosoma cruzi* (the agent of Chagas' disease).[7] Their principal medical impacts are related to itching and inflammation associated with their bites, and emotional and psychological effects on their victims.[8,9]

Adult bed bugs are approximately 5 mm long, oval shaped, and flattened, resembling unfed ticks or small cockroaches (Figure 16.1). Adults are reddish brown (chestnut); immatures look like adults except smaller and yellowish white in color (Figure 16.2). Bed bugs have a pyramidal head with prominent compound eyes, slender antennae, and a long proboscis tucked backward underneath the head and thorax (Figure 16.3). The prothorax (dorsal side, first thoracic segment) has rounded, winglike lateral horns on each side with numerous bristles. Bed bugs can be differentiated from bat bugs by the length of these pronotal bristles.

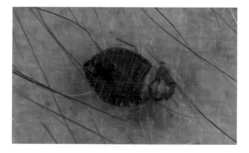

Figure 16.1 Adult bed bug.

Source: Photo copyright 2008 by Jerome Goddard, Ph.D.

Figure 16.2 Bed bug nymph on a dime.

Source: Photo copyright 2020 by Jerome Goddard, Ph.D.

Figure 16.3 The bed bug proboscis is tucked backwards underneath the head and thorax.

Source: Photo copyright 2011 by Jerome Goddard, Ph.D.

Distribution

The common bed bug is cosmopolitan, occurring in temperate regions worldwide. Another human-biting bed bug species, *C. hemipterus*, is also widespread but is mostly found in the tropics or subtropics. Many other bed bug species occur on bats and swallows, but usually do not bite people.[10]

Impact on Human Health

Houses heavily infested with bed bugs may contain literally thousands of specimens under and within the bed and in the mattress seams. There may be black layers of bed bug excrement on the mattress, thousands of cast skins, and eggs several millimeters thick. Infestations like this can lead to anemia in affected individuals due to blood loss.[11] The most common bite reactions are pruritic maculopapular, erythematous lesions at bed bug feeding sites, one per insect (Figure 16.4). Bite lesions may be intensely itchy, but if not abraded, and usually resolve within a week or so.[12–14] Due to hypersensitization, the size and pruritis associated with these common reactions may increase in some individuals who experience repeated bites.[15–17] Some people display more complex cutaneous reactions, including pruritic wheals (local urticaria), papular urticaria, diffuse urticaria,[18–22] and even bullous lesions.[22–25] There are a few reports of systemic reactions from bed bug bites, including asthma, generalized urticaria, and anaphylaxis.[24,26,27] One patient staying in a hotel awakened during the night with severe itching and urticaria on his arm and neck; bed bugs were found in the room.[28] He developed angioedema and hypotension, was hospitalized, and showed transient anterolateral ischemia on electrocardiogram. Eight months later, after an experimental bed bug bite, he developed a wheal at the bite site and generalized itching that required epinephrine to resolve his symptoms.

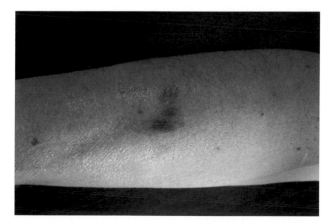

Figure 16.4 Bed bug bites are often red and intensely itchy.

Source: Photo courtesy Dr. Kristine T. Edwards, used with permission

> ### FIVE WAYS BED BUGS MAY AFFECT HUMAN HEALTH
>
> - Psychological or emotional effects from biting (insomnia, anxiety, etc.)
> - Direct effects of bites (itchy, red lesions, etc.)
> - Allergic reactions
> - Anemia (for those living in heavily infested dwellings)
> - Disease transmission (not yet proven)

From the public health perspective, one of the biggest issues associated with bed bugs relates to socioeconomic inequity. Poverty is a driving force in bed bug infestations. Low-income or minority communities have higher unemployment rates and may contain immigrants who have previous exposure to bed bug-infested living conditions. These groups are plagued with persistent bed bug infestations, yet the cost required to rid their homes and apartments of bed bug infestations easily can reach into the thousands. Unfortunately, self-help or do-it-yourself pest control is mostly ineffective and drains what little financial resources these groups have. Accordingly, bed bugs are quickly becoming a pest of poor and underserved communities, and the public health community is in a precarious social justice position if its response (or lack thereof) disproportionately affects underserved populations.[29] Middle- and upper-class communities sometimes become cavalier about this issue, thinking, "Well, that's their problem," not realizing that the bed bug problem will never fully be solved without dealing with pockets of infestations in underserved populations living in low-income or federally subsidized housing units. People living in those circumstances travel, go to libraries, hospitals, schools, and other public places and bring with them bed bugs from home.

Prevention and Control

Protection at a hotel. Travelers should keep in mind that not every hotel room has bed bugs, but some do. Interestingly, bed bugs are just as prone to be found in both low budget and 5-star hotels, so it helps to always be on guard when traveling. Simple precautions may help protect you and your belongings from bed bug infestation. Leave all unnecessary items in your vehicle, such as extra clothing, gear, and equipment. When first entering your hotel room, place luggage on the bathroom vanity until you have had a chance to inspect the premises (do not place luggage on bed, floor, chair, sofa, or luggage rack). Pull back sheets and check mattress and box springs for live bed bugs or black fecal spots. If possible, remove the headboard from the wall and inspect behind it. NOTE: some hotels now are avoiding use of headboards, supposedly to discourage bed bug infestation (Figure 16.5). If any bugs or suspicious signs of infestation are noted in the room, go back to the reception area and request another room.

Protection at home. Keeping bed bugs out of homes can be difficult, especially if homeowners travel a lot or have frequent guests. Furthermore, there are other ways bed bugs may enter your home, such as on used furniture, goods, or items purchased at secondhand stores or garage sales, so those items need to be disinfected (more precisely "dis-insected"). First of all, after traveling, be sure to wash all clothing from luggage in hot water and dry on high heat if possible. Alternatively, place clothing directly in a dryer on high heat. You should seal luggage in plastic bags between uses. It is a good idea to *never* purchase used mattresses, no matter how good a bargain. Other used items brought into the home should be thoroughly inspected for evidence of bed bugs.

Bed bug control. If a bed bug infestation is found in a hotel room, you should immediately report the incident to hotel management. If bed

Figure 16.5 Hotel room with no headboard, a practice often employed as a bed bug prevention measure.

Source: Photo copyright 2021 by Jerome Goddard, Ph.D.

bugs are found in your home, or in used items purchased at thrift stores or garage sales, it is best not to try to spray them yourself with over-the-counter pesticides. Not many insecticides are specifically labeled for bed bugs and might actually make things worse by only partially controlling them. The best thing to do is contact a competent pest exterminator for professional pest service.

References

1. Doggett S, Russell RC. The resurgence of bed bugs, *Cimex* spp., in Australia: experiences from down under. In: Proceed. 6th Inter. Conf. Urban pests, Budapest, Hungary, 13–16 July 2008:1–30.
2. Potter MF. The perfect storm: an extension view on bed bugs. *Am Entomol.* 2006;52:102–104.
3. Potter MF. The business of bed bugs. In: Pest Management Professional Magazine, January issue; 2008:24–29.
4. Potter MF, Rosenberg B, Henriksen M. Bugs without borders: defining the global bed bug resurgence. In: Pest World, Sept/Oct; 2010:8–20.
5. Burton GJ. Bed bugs in relation to transmission of human diseases. *Public Health Rep.* 1963;78:513–524.
6. Goddard J, de Shazo RD. Bed bugs (*Cimex lectularius*) and clinical consequences of their bites. *J Am Med Assoc.* 2009;301:1358–1366.
7. Salazar R, Castillo-Neyra R, Tustin AW, Borrini-Mayori K, Naquira C, Levy MZ. Bed Bugs (*Cimex lectularius*) as vectors of *Trypanosoma cruzi. Am J Trop Med Hyg.* 2015;92(2):331–335.
8. Goddard J, de Shazo RD. Psychological effects of bed bug attacks (*Cimex lectularius* L.). *Am J Med.* 2012;125(1):101–103.
9. Susser SR, Perron S, Fournier M, et al. Mental health effects from urban bed bug infestation (*Cimex lectularius* L.): a cross-sectional study. *BMJ Open.* 2012;2(5).
10. Usinger RL. *Monograph of Cimicidae.* Vol. 7. College Park, MD: Entomological Society of America, Thomas Say Foundation; 1966.
11. Pritchard MJ, Hwang SW. Severe anemia from bed bugs. *CMAJ.* 2009;181(5):287–288.
12. Kemper H. Die Bettwanze und ihre bekampfung. *Schriften uber Hygienische Zoologie, Z Kleintierk Pelztierk.* 1936;12:1–107.
13. Masetti M, Bruschi F. Bedbug infestations recorded in Central Italy. *Parasitol Int.* 2007;56(1):81–83.
14. Ryckman RE. Dermatological reactions to the bites of four species of triatominae (hemiptera: reduviidae) and *Cimex lectularius* L. (hemiptera: cimicidae). *Bull Soc Vector Ecol.* 1985;10:122–125.
15. Bartley JD, Harlan HJ. Bed bug infestation: its control and management. *Mil Med.* 1974;139:884–886.
16. Cestari TF, Martignago BF. Scabies, pediculosis, and stinkbugs: uncommon presentations. *Clin Dermatol.* 2005;23:545–554.
17. Liebold K, Schliemann-Willers S, Wollina U. Disseminated bullous eruption with systemic reaction caused by *Cimex lectularius. J Eur Acad Dermatol Venereol.* 2003;17:461–463.

18. Rook AJ. Papular urticaria. *Ped Clin NA*. 1961;8:817–820.

19. Alexander JO. *Arthropods and Human Skin*. Berlin: Springer-Verlag; 1984.

20. Gbakima AA, Terry BC, Kanja F, Kortequee S, Dukuley I, Sahr F. High prevalence of bedbugs *Cimex hemipterus* and *Cimex lectularis* in camps for internally displaced persons in Freetown, Sierra Leone: a pilot humanitarian investigation. *West Afr J Med*. 2002;21(4):268–271.

21. Brasch J, Schwarz T. [26-year-old male with urticarial papules]. *J Dtsch Dermatol Ges*. 2006;4(12):1077–1079.

22. Cooper DL. Can bedbug bites cause bullous erythema? *J Am Med Assoc*. 1948;138:1206.

23. Hamburger F, Dietrich A. Lichen urticatus exogenes. *Acta Paediat*. 1937;22:420.

24. Kemper H. Beobachtungen ueber den Stech-und Saugakt der Bettwanze und seine Wirkung auf die menschliche Haut. *Zeitschr f Desinfekt*. 1929;21:61–65.

25. Kinnear J. Epidemic of bullous erythema on legs due to bed-bug. *Lancet*. 1948;255:55.

26. Bircher AJ. Systemic immediate allergic reactions to arthropod stings and bites. *Dermatology*. 2005;210(2):119–127.

27. Jimenez-Diaz C, Cuenca BS. Asthma produced by susceptibility to unusual allergens. *J Allergy*. 1935;6:397–403.

28. Parsons DJ. Bed bug bite anaphylaxis misinterpreted as coronary occlusion. *Ohio State Med J*. 1955;51:669.

29. Eddy C, Jones SC. Bed bugs, public health, and social justice, part I, a call to action. *J Environ Health*. 2011;73:8–14.

chapter seventeen

Kissing Bugs

Importance and Physical Description

Many members of the insect family Reduviidae have an elongate (cone-shaped) head, and hence the name conenose bugs (Figure 17.1). A relatively small but important group of these bugs within the subfamily Triatominae feeds exclusively on vertebrate blood. Notorious members of this group are frequently in the genus *Triatoma*, but not all. Triatomines are called kissing bugs because their blood meals are occasionally taken on the face or around human lips (Figure 17.2). Very often, their bites are painless; however, bite reactions may range from a single papule, to giant urticarial lesions, to anaphylaxis, depending on the degree of allergic sensitivity.[1] In fact, at least one death from anaphylaxis due to a kissing bug bite has been reported in the western United States.[2] Kissing bugs may also transmit the agent of Chagas disease, or American trypanosomiasis, one of the most important arthropod-borne diseases in tropical America.

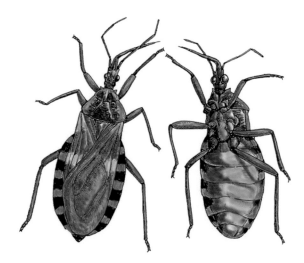

Figure 17.1 Kissing bugs, dorsal and ventral views.

Source: Figure courtesy of Joe MacGown, Mississippi State University, used with permission

249

Kissing bugs are similar in appearance to assassin bugs (other members of the family Reduviidae) and may have orange and black markings where the abdomen extends laterally past the folded wings. The beak is short and three-segmented, and its tip fits into a groove in the venter of the thorax. In addition, the dorsal portion of the first segment of the thorax consists of a conspicuous triangular-shaped structure. Most adult kissing bugs are 1–3 cm long (Figure 17.3) and are good fliers.

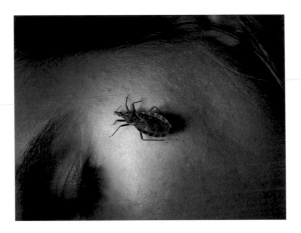

Figure 17.2 Kissing bugs often feed on the face.

Source: Photo copyright 2004 by Jerome Goddard, Ph.D.

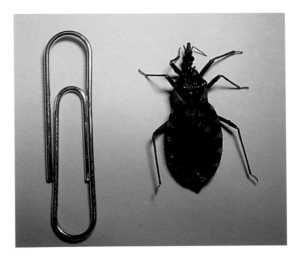

Figure 17.3 Relative size of an adult kissing bug.

Source: Photo copyright 2005 by Jerome Goddard, Ph.D.

Distribution

Kissing bugs occur primarily in the New World (Figure 17.4), although there is at least one species complex (the rubrofasciata complex) reported from port areas throughout the Old World tropics and subtropics, and a few species of the genus *Linshcosteus* occur in India.[3] There are many

Figure 17.4 Approximate geographic distribution of Chagas' disease.

species of kissing bugs that can attack humans, and some are capable of transmitting Chagas disease. The four principal vectors of Chagas disease in Central and South America are *Panstrongylus megistus, Rhodnius prolixus, Triatoma infestans,* and *T. dimidiata. Triatoma gerstaeckeri* and *T. protracta* are important kissing bug species in the southwestern United States, and *T. sanguisuga* occurs throughout much of the southeastern United States.

Impact on Human Health

Kissing bugs feed on a wide variety of small- to medium-sized mammals. One study showed that perhaps they feed on humans more than we know in areas not typically known for their presence.[4] Other than nuisance biting and occasional reports of allergic reactions to their bites, kissing bugs are famous for their vectorial capacity for the agent of Chagas disease, *Trypanosoma cruzi.*[1] Chagas disease is a zoonosis (originally a parasite of wild animals) mostly occurring in Mexico and Central and South America, but a few indigenous cases have been reported in Texas, Tennessee, Louisiana, and California.[5-7] Further, there is serological evidence of *T. cruzi* in dogs, raccoons, opossums, and woodrats as far north as Oklahoma, Virginia, and Maryland.[8,9] Some 8 million people are estimated to be infected with Chagas, with tens of millions of people at risk.[10] Chagas disease has both acute and chronic forms, but is perhaps most well known for chronic sequelae (occurring years to decades later), such as myocardial damage with cardiac dilation, arrhythmias and major conduction abnormalities, and digestive tract involvement such as megaesophagus and megacolon.

Prevention and Control

Kissing bugs are nocturnal insects that seek refuge by day in cracks and crevices of poorly constructed houses or in the loose roof thatching of huts. Personal protection from the bugs involves avoidance, such as not sleeping in thatched-roof houses in endemic areas if possible, and exclusion methods such as bed nets. Also, proper construction of houses, wise choice of building materials, sealing cracks and crevices, and precision targeting of insecticides are all important components of a kissing bug prevention and control plan.

References

1. Moffitt JE, Venarske D, Goddard J, Yates AB, deShazo RD. Allergic reactions to *Triatoma* bites. *Ann Allergy Asthma Immunol.* 2003;91(2):122–128; quiz 128–130, 194.
2. Anonymous. Chamber of Commerce Official's Death Laid to Kissing Bug Bite. In: Los Angeles Times, Los Angeles, CA, September 17, 1990 issue; 1990.

3. Schofield CJ, Galvao C. Classification, evolution, and species groups within the Triatominae. *Acta Trop.* 2009;110:88–100.

4. Klotz SA, Schmidt JO, Dorn PL, Ivanyi C, Sullivan KR, Stevens L. Free-roaming kissing bugs, vectors of Chagas' disease, feed often on humans in the Southwest. *Am J Med.* 2014;127:421–426.

5. Harwood RF, James MT. *Entomology in Human and Animal Health.* 7th ed. New York: Macmillan; 1979.

6. Herwaldt BL, Grijalva MJ, Newsome AL, et al. Use of PCR to diagnose the fifth reported U.S. case of autothonous transmission of *Trypanosoma cruzi* in Tennessee. *J Infect Dis.* 1998;181:395–399.

7. Schiffler RJ, Mansur GP, Navin TR, Limpakarnjanarat K. Indigenous Chagas disease in California. *J Am Med Assoc.* 1984;251:2983–2984.

8. Bradley KK, Bergman DK, Woods JP, Crutcher JM, Kirchoff LV. Prevalence of American trypanosomiasis among dogs in Oklahoma. *J Am Vet Med Assoc.* 2000;217:1853–1857.

9. Kribs-Zaleta C. Estimating contact process saturation in sylvatic transmission of *Trypanosma cruzi* in the United States. *PLoS Negl Trop Dis.* 2010;4:e656.

10. Kirchhoff LV. American trypanosomiasis (Chagas' disease). In: Guerrant RL, Walker DH, Weller PF, eds. *Tropical Infectious Diseases: Principles, Pathogens, and Practice.* 3rd ed. New York: Saunders (Elsevier); 2011:689–695.

chapter eighteen

Mites

Importance and Physical Description

Chiggers

Larval stage mites in the family Trombiculidae, sometimes called *chiggers* (Figure 18.1a), *harvest mites*, or *red bugs*, are medically important pests around the world, primarily because they cause dermatitis and may transmit the agent of scrub typhus. Adult chiggers are oval shaped (approximately 1 mm long) with a bright red, velvety appearance, but it is only the larval stage that attacks vertebrate hosts. Chigger larvae are very tiny (0.2 mm long), round mites with numerous setae (Figure 18.2). The mites may be red, yellow, or orange in color and have a single dorsal plate (scutum) bearing two sensillae and four to six setae. Identification to the species level is extremely difficult and expert technical help is required (see Chapter 8).

Other Biting Mites

There are actually only two species of true human parasitic mites—scabies mites and hair follicle mites—which can permanently live on humans. As far as is known, follicle mites cause no pathology, but scabies mites

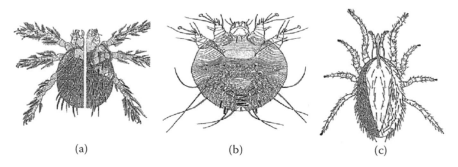

(a)   (b)   (c)

Figure 18.1 Various human-biting mites: (A) chigger, (B) scabies mite, and (C) tropical rat mite.

Source: From the U.S. Food and Drug Administration

Figure 18.2 Larval chigger mite, the stage that bites humans.

Source: From the U.S. Department of Agriculture, photo by Ron Ochoa

cause significant morbidity, infesting millions of people worldwide annually.[1,2] Scabies mites are extremely tiny (0.2–0.4 mm long), oval, eyeless mites with stubby legs (Figure 18.1b). They cause linear burrows on the skin containing the mites and their eggs, which leads to sensitization and subsequent itching. Besides follicle and scabies mites, many other species of mites sometimes may bite people (but do not live on them) and are important to public health, such as tropical rat mites, tropical fowl mites, northern fowl mites, chicken mites, and house mouse mites. Most are important only as causes of itch or dermatitis, but the house mouse mite, *Liponyssoides* (formerly *Allodermanyssus*) *sanguineus*, can be a vector of rickettsialpox in the eastern United States.[3]

Tropical rat mites are large and easier to recognize than many species (Figure 18.1c). Females have scissorlike chelicerae, narrow, tapering dorsal and genitoventral plates, and an egg-shaped anal plate. In addition, the protonymphal stage and adult females suck blood, and become tremendously distended after feeding. They look like tiny engorged ticks. Tropical fowl mites and northern fowl mites are similar in appearance to the tropical rat mite, except tropical fowl mites have a wider dorsal plate, and the northern fowl mite has a much shorter sternal plate. The chicken mite has large dorsal and anal plates, a short sternal plate, and long, needle-like chelicerae. The house mouse mite can be distinguished from most other mites by the presence of two dorsal shields, a large anterior plate, and a small posterior plate bearing one pair of setae.

House Dust Mites

Several species of mites have been reported from house dust, but only three are major sources of mite allergen, and thus medically important: *Dermatophagoides farinae*, *D. pteronyssinus*, and *Euroglyphus maynei*. Considerable research on house dust allergy has revealed that both *D. pteronyssinus* and *D. farinae* (the most familiar house dust mites) possess powerful allergens in the mites themselves, as well as in their secretions and excreta.

Figure 18.3 House dust mites in sample.

Source: Photo copyright 2009 by Jerome Goddard, Ph.D.

Fecal pellets seem to be especially allergenic. Adult house dust mites are white to light tan in color, about 0.5 mm long, and with a cuticle showing numerous fine striations. The mites have plump bodies (not flattened) (Figure 18.3), well-developed chelicerae, and suckers at the ends of their tarsi.

Distribution

Chiggers

Although chigger mites may occur from Alaska to New Zealand, species are most common in tropical and subtropical areas. Some particularly common pest chiggers are *Eutrombicula alfreddugesi* in the United States and parts of Central and South America, *Neotrombicula autumnalis* in Europe, and *E. sarcina* in Asia and Australia. Vectors of the scrub typhus agent in Japan, Southeast Asia, and parts of Australia are in the genus *Leptotrombidium*.

Other Biting Mites

The tropical rat mite is found on all continents in association with rats; the tropical fowl mite is found in many areas of the world, but especially the tropics; the northern fowl mite occurs in temperate regions worldwide; and the house mouse mite occurs in Northern Africa, Asia, Europe, and the United States (especially the northeast).

House Dust Mites

House dust mites may be found in human habitations worldwide.

Impact on Human Health

Chiggers

Infestation with trombiculid larvae is called *trombiculosis*, or sometimes *trombidiosis*. Larval chiggers crawl up on blades of grass or leaves and subsequently get on passing vertebrate hosts. On humans, they attach to the skin anywhere (Figure 18.4), but particularly where clothing fits snugly or where flesh is tender, such as ankles, groin, or waistline. Chiggers then attach to the skin with their mouthparts, inject saliva into the wound (which dissolves tissue), and then suck up this semidigested material. They do not actually burrow in human skin; only the chelicerae penetrate the skin of the host. Feeding is aided by formation of a stylostome, or feeding tube, created by interaction of the saliva and surrounding skin tissue. *Scrub typhus*, a zoonotic rickettsial infection caused by *Orientia tsutsugamushi*, is mite-borne and occurs over much of Southeast Asia, India, Sri Lanka, Pakistan, islands of the southwest Pacific, and coastal Australia (Queensland) (Figure 18.5). Chiggers are the vectors, so the name chigger-borne ricketsiosis might be more appropriate for the disease. Scrub typhus occurs in nature in small, but intense foci of infected host animals. These "mite islands" or "typhus islands" occur where the appropriate combination of rickettsiae, vectors, and suitable animal hosts occurs.[4,5] Epidemics occur when susceptible individuals come into contact with these areas. Historically, military operations have often been severely affected by scrub typhus.[6]

Other Biting Mites

Scabies causes much misery and suffering among humans worldwide every year, and has done so since ancient times. Furthermore, in persons

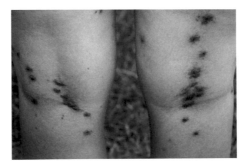

Figure 18.4 Chigger bites on backside of legs.

Source: Photo courtesy Dr. Wendy C. Varnado

Figure 18.5 Approximate geographic distribution of scrub typhus.

with compromised immune systems, the mites may multiply uncontrollably, leading to a condition called *crusted scabies* or *Norwegian scabies*. Most biting mites such as scabies or tropical rat mites only cause itching and dermatitis; however, at least one group, the chiggers, may transmit a disease agent in parts of Asia (see the previous paragraph). In the United States, *rickettsialpox*, carried by the house mouse mite, is a relatively mild bacterial infection caused by *Rickettsia akari*, a member of the spotted fever group rickettsiae. The disease often begins with a nonitchy red papule where the infected mite bit the patient. This may further develop into an ulcer-like sore that turns into a brown or black scab called an eschar. Fever, headache, malaise, and myalgias accompany the signs at the bite site(s). The disease is relatively rare, with only about 1,000 cases ever reported.[7]

SOME COMMON HUMAN-BITING MITES

- Tropical rat mites
- Tropical fowl mites
- Scabies mites
- Chigger mites
- Straw itch mites
- Various stored food products mites (grain mites, etc.)

House Dust Mites

A long time ago, some people were said to be allergic to house dust, but in the 1960s, Dutch researchers realized that *house dust allergy* was really due to house dust mites.[8,9] House dust mites commonly infest homes throughout much of the world and feed on shed human skin scales, mold, pollen, feathers, and animal dander. They are barely visible to the naked eye and live most commonly in mattresses, chairs, and other furniture where people spend a lot of time. Dust mites are not poisonous and do not bite or sting, but they contain powerful allergens in their excreta, exoskeleton, and scales. House dust mites are the most important domestic source of allergic disease such as allergic rhinitis, asthma, and atopic dermatitis.[10,11] In fact, much asthma is due to dust mites; for example, Htut and Vickers[12] say that house mites are the major cause of asthma in the UK.

Prevention and Control of Mites

The main strategy for prevention and control of mites is to identify the offending species and either: (1) eliminate the breeding site/source, or (2) avoid exposure. For example, chigger mites can be avoided by reducing outdoor activity in infested areas during seasonal peaks, or wearing boots and repellents.[13] Infestations with scabies can be prevented by limiting exposure to infested persons or animals. If infested, patients can be treated with a variety of drugs or pesticidal creams and lotions.[14] House dust mite allergy is managed by immunotherapy using mite extracts and by efforts to minimize the level of dust mites in the patient's home.

References

1. Mellanby K. *Scabies*. Oxford: Oxford University Press; 1943:81 pp.
2. Service MW. *Medical Entomology for Students*. 5th ed. Cambridge, UK: Cambridge University Press; 2012.
3. Huebner RJ, Jellison WL, Pomerantz C. Rickettsialpox—a newly recognized rickettsial disease. IV. Isolation of a rickettsia apparently identical with the causative agent of rickettsialpox, from *Allodermanyssus sanguineus*, a rodent mite. *Public Health Rep.* 1946;61:1677–1682.
4. Heymann DL, ed. *Control of Communicable Diseases Manual*. 20th ed. Washington, DC: American Public Health Association; 2015.
5. Varma MGR. Ticks and mites. In: Lane RP, Crosskey RW, eds. *Medical Insects and Arachnids*. London: Chapman and Hall; 1993:chap. 18.
6. Cushing E. *History of Entomology in World War II*. Washington, DC: Smithsonian Institution; 1957.
7. Mullen GR, O'Connor BM. Mites (Acari). In: Mullen GR, Durden LA, eds. *Medical and Veterinary Entomology*. New York: Elsevier; 2009:433–492.
8. Spieksma FTM. The mite fauna of house dust, with particular reference to the house dust mite. *Acarologia*. 1967;9:226–234.

9. Spieksma FTM. The house dust mite, *Dermatophagoides pteronyssinus*, producer of house dust allergen. Thesis, University of Leiden, Netherlands; 1967:65 pp.

10. Cameron MM. Can house dust mite-triggered atopic dermatitis be alleviated using acaricides. *Br J Dermatol*. 1997;137:1–8.

11. Huang FL, Liao EC, Yu SJ. House dust mite allergy: its innate immune response and immunotherapy. *Immunobiology*. 2018;223(3):300–302.

12. Htut T, Vickers L. The prevention of mite-allergic asthma. *Inter J Environ Health Res*. 1995;5:47–61.

13. Breeden GC, Schreck CE, Sorensen AL. Permethrin as a clothing treatment for personal protection against chigger mites. *Am J Trop Med Hyg*. 1982;33:589–592.

14. Bope ET, Kellerman R. *Conn's Current Therapy*. Philadelphia: Elsevier Saunders; 2017.

chapter nineteen

Pests Involved in Mechanical Disease Transmission

Background and Medical Significance

A wide variety of arthropod and vertebrate pests may be involved in mechanical transmission of disease agents, including ants, beetles, cockroaches, flies, and rodents. Mechanical transmission of disease agents occurs when arthropods physically transport pathogens from one place or host to another via body parts. For example, ants, flies, and cockroaches have numerous hairs, spines, and setae on their bodies that may pick up contaminants as the insects feed on dead animals or excrement. When they subsequently walk on food, food preparation surfaces, or other sensitive areas, mechanical transmission can occur.[1-3] In modern society we tend to consider this information as a "given," something like, "duh, of course everyone knows that." But just 150 years ago, flies were regarded as harmless nuisances, perhaps even beneficial.[4] We now know mechanical transmission of disease agents is a significant public health issue. For example, I have seen cockroaches emerging from floor drains in a nursing home kitchen and also ants crawling around on IV tubes in hospital rooms. Recently, I saw flies on a toilet seat in the bathroom of a major commercial restaurant chain (Figure 19.1). Health department food inspectors often say that restrooms of a restaurant mirror the kitchen, so how would you like to eat at that one?

Mechanical transmission can also occur if a bloodfeeding arthropod has its feeding event disrupted. For example, if a mosquito feeds briefly on a viremic bird and is interrupted, subsequent feeding on a second bird could result in virus transmission. This would be similar to an accidental needlestick. The main point about mechanical transmission is that the pathogen undergoes no development (cyclical changes in form and so forth) and no significant multiplication in the vector.

Filth Flies

Flies in the families Muscidae (house flies), Calliphoridae (blow flies), and Sarcophagidae (flesh flies) are often termed filth flies because of

Figure 19.1 Flies on a toilet seat in a major restaurant chain.

Source: Photo copyright 2019 by Jerome Goddard, Ph.D.

their unsanitary breeding and feeding habits. For example, house flies, *Musca domestica*, regurgitate liquid through their proboscis as they probe around on various surfaces, and also may deposit fecal matter as they crawl over food sources. House flies occur worldwide in association with human dwellings and are about 5 to 8 mm long with a dull gray thorax and abdomen (not shiny) (Figure 19.2). The thorax has four longitudinal dark stripes, and there are pale yellow areas on each side of the abdomen. Mature house fly larvae are 10 to 13 mm long and usually creamy white in color (Figure 19.3). Overall, the larvae have a conical shape very much like a carrot with two dark-colored mouth hooks at the narrow end and two oval spiracular plates or openings at the broad posterior end. There are no soft protuberances surrounding the spiracular openings in larval house flies. In addition, the three slits in their posterior spiracular plates of larvae are curvy as opposed to straight, which is a diagnostic feature for this species.

Figure 19.2 House fly adult.

Source: Photo copyright 2011 by Jerome Goddard, Ph.D.

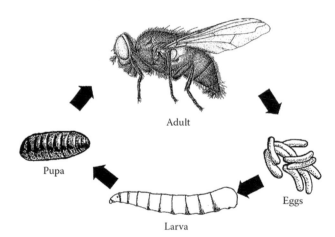

Adult

Pupa

Eggs

Larva

Figure 19.3 House fly life cycle.

Source: From the U.S. Department of Agriculture, Agricultural Handbook No. 655, 1991

SOME EXAMPLES OF MECHANICAL DISEASE TRANSMISSION

- Cockroaches in restaurants or food preparation areas
- Ants on IV tubes in hospitals
- Flies breeding in excrement or garbage and later entering homes, restaurants, or healthcare facilities
- Rodents walking across food preparation surfaces

Various species of flesh flies occur worldwide. One of the most common is *Sarcophaga hemorrhoidalis*, which is virtually worldwide in distribution and common in the United States. This species has a red-tipped abdomen (hence the name hemorroidalis). Flesh flies look like house flies but are generally larger (11–13 mm long); they have three dark longitudinal stripes on their thorax, a checkerboard pattern of gray on the abdomen, and sometimes a reddish brown tip on the abdomen (Figure 19.4). The larvae of flesh flies are similar to those of house flies, except they have a ring of small protuberances surrounding the blunt end, straight spiracular slits, and an incomplete ring around the spiracular plate.

Blow flies (also known as green or bluebottle flies) are about the same size as flesh flies, although some of the bluebottle flies (genus *Calliphora*) are larger and more robust. Blow flies, with a few exceptions, are metallic bronze, green, black, purplish, or blue colored. They are the commonly encountered "green flies" seen on flowers, dead animals, or feces, or occasionally indoors (Figure 19.5). These green flies are usually in the genus *Lucilia*[5] (they were formerly classified as *Phaenicia*).[6] One common blow fly, *Lucilia sericata*, is a cosmopolitan pest. *Calliphora vicina* is one of the most common bluebottle species in Europe and North America. Blow fly maggots resemble both house fly and flesh fly maggots but have the ring of protuberances on the blunt end (Figure 19.6), straight spiracular slits, and often a complete sclerotized ring around the spiracular plate (this feature is variable).

Figure 19.4 Sarcophaga hemorrhoidalis, a flesh fly.

Figure 19.5 Typical blow fly or green fly.

Source: Courtesy of Dr. Blake Layton, Mississippi State University Extension Service.

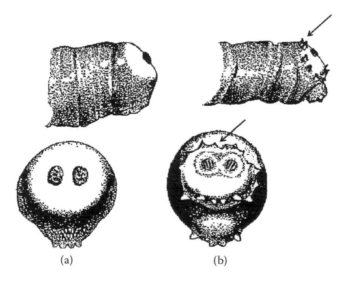

(a)   (b)

Figure 19.6 (a) House fly and (b) blow fly larvae, showing differences.

Rodents

Domestic rodents such as brown rats, roof rats, and house mice are called commensal rodents and may be involved in mechanical disease transmission as well. The brown rat or sewer rat is large bodied with a long, sparsely haired and scaly tail. Although it may both swim and climb, this species prefers terrestrial locations, constructing extensive underground burrows with branching tunnels and exits. Brown rats (also called Norway rats; Figure 19.7) are commonly found in urban areas, around ditches, garbage receptacles, sewers, and human dwellings. Essentially, any area occupied by humans can be a habitat for the brown rat. On the other hand, roof rats, also known as black rats or ship rats, are medium sized with a longer hairless tail and larger ears. They are more agile than brown rats and may run between buildings on telephone or other wires. Roof rats prefer to nest in dry areas aboveground in trees or other elevated locations. The roof rat is especially common in coastal areas. Lastly, the house mouse is considerably smaller than the rats listed above, only being about 5 in. long. Fur color may range from grayish brown to pale gray. The house mouse is found worldwide wherever humans are, and is commonly found in houses, granaries, and barns. They often live and breed under floors and in wall voids of human habitations. Although they can swim, house mice generally avoid water and damp conditions.

Cockroaches

Cockroaches are well-known mechanical transmitters of disease agents. They are dorsoventrally flattened, fast-running insects that generally live in warm, moist, secluded areas. Cockroaches have prominent, multisegmented filiform antennae, cerci on the abdomen, and two pairs of

Figure 19.7 Norway rat.

Source: Courtesy Stoy Hedges Pest Consulting, used with permission

wings. The front wings are typically hardened and translucent, whereas hind wings are membranous. In some species, wings are rudimentary or absent. Cockroaches are variously colored, with most domestic species being reddish brown, brown, or black, while the Cuban cockroach is bright green. Many species can fly, but the domestic U.S. species rarely do so; however, the imported Asian cockroach in the southeastern United States can fly and may come to lights.[7] Adult German and brown-banded cockroaches are approximately 15 mm long (Figure 19.8), whereas the American and oriental cockroaches are 30–50 mm long (Figure 19.9).

Figure 19.8 Adult German cockroaches showing two dark lines on the pronotum, a diagnostic feature.

Source: Photo copyright 2019 by Jerome Goddard, Ph.D.

Figure 19.9 American cockroach.

Source: Photo copyright 2005 by Jerome Goddard, Ph.D.

Some tropical species are even bigger. Immature cockroaches look similar to adults (except they have no wings), and some of the first nymphal stages are so small as to be confused with ants. Cockroaches may adversely affect human health in several ways: biting feebly, especially gnawing the fingernails of sleeping children; entering human ear canals; contaminating food and imparting an unpleasant odor and taste; and transmission of disease organisms mechanically on their body parts.[8-10] In fact, pathogenic bacteria and fungi have been found on cockroaches. For example, Burgess[11] reported a strain of *Shigella dysenteriae* from German cockroaches that was responsible for a disease outbreak in Northern Ireland.

Contributing factors for mechanical transmission of disease agents include anything that aids the juxtaposition of the afore-mentioned pests with food production/consumption and wastes or other unsanitary conditions. Flies, rodents, and cockroaches by their indiscriminant movements from unsanitary places to human food mechanically transmit disease-producing organisms to humans. Of course, subsequent cleaning/processing/cooking kills most of the microorganisms transmitted in this manner, or the vectors themselves are washed away or killed at some later stage of processing. However, the foundational aspects of sanitation are improved by preventive measures that keep mechanical transmission to a bare minimum. Uncovered and unrefrigerated foods also contribute to mechanical disease transmission by pests. Intuitively, more pests equal higher chances of contamination, so lack of appropriate pest control is a contributing factor.

Prevention, Treatment, and Control

The most important control measure for any of these insect or rodent mechanical transmitters of disease agents is to find and eliminate their breeding sites. Dr. L. O, Howard, both a physician and an entomologist said this about house fly control over 100 years ago, "Killing a few hundred, or a few thousand, or a few hundred thousand flies with a fly swatter will do little good if their breeding places are left undisturbed." [4] Secondly, protecting food and food preparation surfaces from arthropod and rodent pests is important to prevent mechanical transmission of diseases. *Prevention* and *exclusion* of pests should be emphasized. No matter how new and "sealed" a food or healthcare establishment is, pests may still enter the establishment if doors or windows are left open (Figure 19.10). Once inside, flies, ants, and cockroaches may pick up germs on contaminated surfaces and transfer them to sensitive areas (Figure 19.1). Covering and refrigerating foods is helpful in preventing mechanical transmission of pathogens.

Figure 19.10 Door of food establishment propped open.

Source: Photo copyright 2018 by Jerome Goddard, Ph.D.

IPM and Alternative Control Methods

Any systematic and well-run pest control program will include integrated pest management (IPM) techniques. Nonchemical pest control methods such as air (fly) curtains, heat, and cold should be utilized, as well as chemical products such as baits, dusts, and sprays, all used only according to their labels. A good policy for food establishments is to use the least amount of the least toxic products for pest control. Interestingly, not all states require retail food establishments to have a licensed pest management professional for their pest control services. The restaurant owners may do it themselves with pesticides and products purchased over the counter. Certainly, this is worrisome, but it is entirely legal in some states as long as the pesticide product labels are followed precisely.

References

1. Bressler K, Shelton C. Ear foreign-body removal: a review of 98 consecutive cases. *Laryngoscope*. 1993;103:367–370.
2. Kopanic RJ, Sheldon BW, Wright CG. Cockroaches as vectors of *Salmonella*: laboratory and field trials. *J Food Prot*. 1994;57:125–132.
3. Zurek L, Schal C. Evaluation of the German cockroach as a vector for verotoxigenic *Escherichia coli* F18 in confined swine production. *Vet Microbiol*. 2004;101:263–267.
4. Howard LO. A fifty year sketch history of medical entomology and its relation to public health. In: Ravenel MP, ed. *A Half Century of Public Health*. New York: American Public Health Association; 1921:412–438.
5. Whitworth T. Keys to the genera and species of blow flies of America North of Mexico. *Proc Ent Soc Washington*. 2006;108:689–725.
6. Hall DG. *The Blow flies of North America*. Washington, DC: Thomas Say Foundation; 1948.
7. Brenner RJ, Koehler PG, Patterson RS. The Asian cockroach. *Pest Manag*. 1986;5:17–22.
8. Brenner RJ, Koehler PG, Patterson RS. Health implications of cockroach infestations. *Infect Med*. 1987;4:349–360.
9. Lemos AA, Lemos JA, Prado MA, et al. Cockroaches as carriers of fungi of medical importance. *Mycoses*. 2006;49(1):23–25.
10. Tatfeng YM, Usuanlele MU, Orukpe A, et al. Mechanical transmission of pathogenic organisms: the role of cockroaches. *J Vector Borne Dis*. 2005;42(4):129–134.
11. Burgess N. Biological features of cockroaches and their sanitary importance. In: Bajomi, D. and Erdos, G., eds. Lectures Delivered at the International Symposium on Modern Defensive Approaches to Cockroach Control. Budapest, Hungary: The Public Health Commission; 1982:45–50.

chapter twenty

Arthropod Bites or Stings

Introduction and Medical Significance

Biting and stinging arthropods in the urban and suburban environment may negatively affect human health by the lesions they produce as well as allergic reactions.[1-8] Skin lesions resulting from arthropod exposure have various pathologic origins, such as direct damage to tissue, allergic reactions to venom or saliva, or infectious diseases. Allergic reactions aside, much human morbidity results from direct injury of arthropod biting/stinging. Direct injury can occur from mouthparts, stingers, or urticating hairs piercing human skin,[9] or from proteins in venom or saliva causing direct mast cell degranulation, leading to allergic reactions.[10] In addition, scratching may lead to secondary infections, wherein strep or staph bacteria are rubbed into the skin via the bite/sting punctum (Figure 20.1). For more information, the reader should consult the *Clinician's Guide to Common Bites and Stings*.[11]

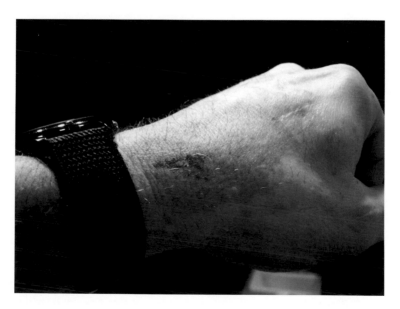

Figure 20.1 Bite or sting lesions can be scratched and become secondarily infected.

Pathogenesis

Bite Apparatus

Bites from arthropods may be defensive only or for purposely taking a bloodmeal. Mosquitoes, ticks, fleas, lice, and other bloodsucking insects are well adapted for blood feeding (Figure 20.2A–20.2C). In medical entomology, insect mouthparts are generally divided into three broad categories: (1) biting and chewing, (2) sponging, and (3) piercing-sucking. There are numerous adaptations or specializations within these mouthpart categories among the various insect orders (Figure 20.3 shows variations among the Diptera). Biting and chewing mouthpart types, such as those in food pest insects, and sponging mouthpart types, found in filth flies, are of little significance regarding human bites, but piercing-sucking mouthparts, and especially the bloodsucking types, are considerably important. There is some variability in insect piercing-sucking mouthparts, mainly in

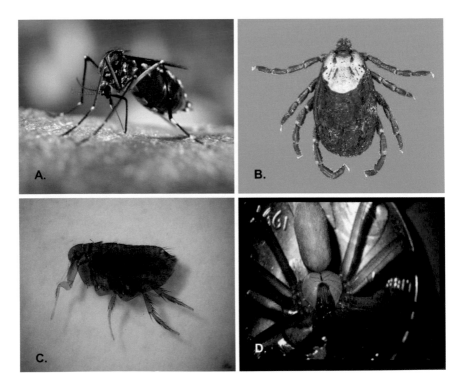

Figure 20.2 Various biting arthropods.

Source: Photo credits: (A) From the Centers for Disease Control; (B) From Dr. Blake Layton, Mississippi State University; (C) Copyright Jerome Goddard, Ph.D.; and (D) From Dr. Barry Engber, formerly North Carolina Department of Environment and Natural Resources

Typical Mouthparts of
Medically Important Diptera

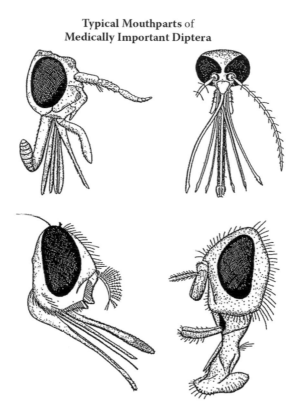

Figure 20.3 Typical mouthparts of medically important Diptera.

Source: From U.S. Navy Laboratory Guide to Medical Entomology, 1943

the number and arrangement of needle-like blades (stylets) and the shape and position of the lower lip of insect mouthparts, the labium. Actually, the proboscis of an insect with piercing-sucking mouthparts is an ensheathment of the labrum, stylets, and labium. These mouthparts are positioned in a way that they form two tubes. One tube is narrow, being a hollow pathway along the hypopharynx, and the other is wider, formed from the relative positions of the mandibles or maxillae. During biting, saliva flows down the narrow tube entering the wound, and blood returns through the wider tube by action of the cibarial or pharyngeal pump.

Sting Apparatus

In the social Hymenoptera, bees, wasps, and some ants may sting (Table 20.1, Figure 20.4A–20.4C). Only the queen or other reproductive caste member lays eggs; the workers gather food, conduct other tasks, and sting

Table 20.1 Some Ants That Sting.

Common name	Scientific name	Geographic distribution
Bullet ants	Subfamily Pomerinae	South America
Army ants	New World, subfamily Ecitoninae, Old World species mostly in tribes Aenictini and Dorylini	Tropics, but primarily Africa and South America
Bulldog ants	*Myrmecia* spp.	Australia
Fire ants (imported and native)	*Solenopsis* spp.	Worldwide
European fire ants	*Myrmica* spp.	Europe and North America
Harvester ants	*Pogonomyrmex* spp.	North and South America
Pavement ants	*Tetramorium* spp.	Worldwide
Twig ants	*Pseudomyrmex* spp.	North America

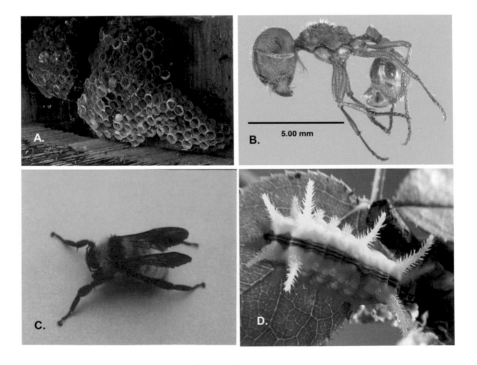

Figure 20.4 Various stinging arthropods.

Source: Photo credits: (A, C, D) Copyright Jerome Goddard, Ph.D., (B) From Joe MacGown, Mississippi State University)

intruders. This is because in stinging wasps, bees, and ants, the stinger is a modified ovipositor, or egg-laying device, that usually no longer functions in egg laying. A normal, egg-laying ovipositor consists of three pairs of elongate structures, called valves, which can insert the eggs into plant tissues, soil, and so forth. One pair of the valves makes up a sheath and is not a piercing structure, whereas the other two pairs form a hollow shaft that can pierce substrate in order for the eggs to pass down through. Two accessory glands within the body of the female inject secretions through the ovipositor to coat the eggs with a glue-like substance.

For an ovipositor modified to enable stinging, the genital opening from which the eggs pass is anterior to the sting apparatus, which is flexed up out of the way during egg laying. The accessory glands have also been modified. One is now a venom gland and the other, called the Dufour's gland, may be involved in production of pheromones. The venom gland is connected to a venom reservoir containing up to 0.1 ml of venom in some of the larger wasp species.

The stinger works well for piercing vertebrate skin. In yellowjackets, there are two lancets and a median stylet that can be extended and thrust into a victim's skin. They do not penetrate skin in one single stroke, but instead, by alternating forward strokes of the lancets sliding along the shaft of the stylet. The tips of the lancets are slightly barbed (even recurved like a fishhook in honey bees) so that they are essentially sawing their way through the victim's skin. Venom is injected by contraction of the venom sac muscles and flows through a channel formed by the lancets and shaft. Honey bees have barbed lancets that hold the stinger in place, preventing it from being withdrawn from vertebrate skin. Thus, the sting apparatus is torn out as the bee flies away. Other hymenopterans, on the other hand, can sting repeatedly.

In addition to wasps, ants, and bees, scorpions can also sting (Figure 20.5), and various moth larvae (caterpillars) possess urticating hairs, setae, or

Figure 20.5 Typical scorpion.

Source: Photo courtesy Stoy Hedges Pest Consulting, used with permission

spines that may secrete venom when exposed to human skin (Figure 20.4D). These urticating caterpillars may cause severe burning, swelling, numbness, urticaria, and even intense stabbing pain.[12-14] More severe cases with systemic symptoms have been reported, including nausea, vomiting, paralysis, renal failure, and anaphylactic shock.[15,16]

Direct Damage to Tissue

Direct tissue damage from stings or bites may lead to development of skin lesions. Arthropod mouthparts puncture skin in various ways (through a siphoning tube, scissor-like blades, and so on), leading to skin damage; hence damage may be a small punctum, dual puncta (from fangs), or lacerations. However, most lesions result from host immune reactions to salivary secretions or venom. Arthropod saliva is important during feeding in order to lubricate the mouthparts on insertion, increase blood flow to the bite site, inhibit coagulation of host blood, anesthetize the bite site, suppress the host's immune and inflammatory responses, and aid in digestion. In contrast, venom from certain spiders may directly cause tissue death (necrosis) in human skin. In the United States violin spiders (Figure 20.2D) are primarily responsible for necrotic skin lesions, although sac spiders (*Cheiracanthium* spp.) and hobo spiders are sometimes reported to cause necrotic arachnidism.[17,18] Brown recluse spider venom contains a lipase enzyme, sphingomyelinase D, which is the primary necrotic agent involved in the formation of the typical lesions. Neutrophil chemotaxis may be induced by sphingomyelinase D. The influx of neutrophils into the area contributes to the formation of the necrotic lesion.

Infectious Complications

Infection with common bacterial pathogens can occur secondarily at any place where skin integrity is disrupted, whether by necrosis, puncture, or excoriation (Figure 20.1).[19] Infection may lead to cellulitis, impetigo, ecthyma, folliculitis, furunculosis, and other manifestations. Three clinical findings may be helpful in making a diagnosis of secondary bacterial infection: (1) increasing erythema, edema, or tenderness beyond an anticipated pattern of response; (2) regional lymphadenopathy, but keep in mind that it may also be present in response to the primary lesion without infection; and (3) lymphangitis suggesting streptococcal involvement.

Clues to Recognizing Insect Bites or Stings

Differential Diagnosis

Physicians and other healthcare providers should keep in mind that it is extremely difficult to identify the causative arthropod from a lesion alone. As for stings, Alexander[1] described a typical hymenopteran sting (excluding ants) as a central white spot marking the actual sting site surrounded by an erythematous halo. The entire lesion is rarely more than a few square centimeters in diameter. Certainly, allergic reactions may lead to much larger lesions. Alexander also described an initial rapid dermal edema resulting from stings with neutrophil and lymphocyte infiltration. Other cells, such as plasma cells, eosinophils, and histiocytes, appeared later.

DIAGNOSIS OF INSECT BITES OR STINGS DEPENDS ON

1. Maintaining a proper index of suspicion in this direction (especially during the summer months)
2. A familiarity of the insect fauna in one's area
3. Obtaining a good history

In any patient complaining of itching, arthropod bites should be considered in the differential diagnosis. Bites, as opposed to stings, are characterized by urticarial wheals, papules, vesicles, and less commonly, blisters. After a few days or even weeks, secondary infection, discoloration, scarring, papules, or nodules may develop at the bite site. Complicating the clinical picture is development of late cutaneous allergic responses (delayed hypersensitivity) in some atopic individuals. Diagnosis based on biopsies of papules or nodules may be extremely difficult, which may reveal a dense infiltrate of a mixture of inflammatory cells, such as lymphocytes, plasma cells, histiocytes, giant cells, neutrophils, and eosinophils. Eosinophils are commonly seen in skin samples from arthropod bites. There can be a dense infiltration of neutrophils, resembling an abscess. Occasionally pieces of arthropod mouthparts may still be seen within the lesion, and there may be a granulomatous inflammation in and around these mouthparts. However, even in light of the dermatopathological findings associated with various insects (described previously), sometimes all that can be said about a lesion is "likely an arthropod assault."

To aid in diagnosis, it is very important to find out what the patient has been doing lately, for example, hiking, fishing, gardening, cleaning out a shed, and so forth. It is also helpful to ascertain if brown recluse or black widow spiders have ever been seen on the patient's premises. However, even history can be misleading in that patients may present a lesion that they think is a bite or sting, when in reality the correct diagnosis is a staph infection, folliculitis, or delusions of parasitosis. Physicians need to be careful not to diagnose "insect bites" based on lesions alone and should call on entomologists to examine samples.

Conclusions

The skin is a human's first line of defense against invasion or external stimuli, and it may react in a variety of ways against all kinds of stimuli— physical or chemical—including arthropods and their emanations. Lesions may result from arthropod exposure, although not all lesions have the same pathological origin—some are due to mechanical trauma, some due to infectious disease processes, and some result from hypersensitization processes. Physicians and other healthcare providers are frequently confronted with patients claiming that their skin lesions are due to a mysterious arthropod bite or sting. Diagnosis in such cases is difficult, but may be aided by asking the patient key questions about the event and any recent activity leading to arthropod exposure. The following kind of questions may provide useful information: Did you see the thing that bit or stung you? Was it wormlike? Did it fly away? Most treatments for skin lesions are symptomatic only (except in cases of infectious diseases) and involve counteracting immune responses to venoms, salivary secretions, or body parts using various combinations of antihistamines and corticosteroids.

References

1. Alexander JO. *Arthropods and Human Skin*. Berlin: Springer-Verlag; 1984.
2. Allington HV, Allington RR. Insect bites. *J Am Med Assoc*. 1954;155:240–247.
3. Frazier CA. Diagnosis of bites and stings. *Cutis*. 1968;4:845–849.
4. Goddard J. Arthropods, tongue worms, leeches, and arthropod-borne diseases. In: Guerrant RL, Walker DH, Weller PF, eds. *Tropical Infectious Diseases: Principles, Pathogens, and Practice*. 3rd ed. Philadelphia: Elsevier Saunders; 2011:868–878.
5. Moffitt JE. Reactions to insect bites and stings: what about the orphan insects? *Ann Allergy Asthma Immunol*. 2004;93(6):507–509.
6. Moffitt JE, de Shazo RD. Allergic and other reactions to insects. In: Rich RR, Fleisher WT, Kotzin BL, Schroeser HW, eds. *Rich's Clinical Immunology: Principles and Practice*. 2nd ed. New York: Mosby; 2001.
7. Moraru GM, Goddard JI. *The Goddard Guide to Arthropods of Medical Importance*. Boca Raton, FL: CRC Press; 2019.
8. O'Neil ME, Mack KA, Gilchrist J. Epidemiology of non-canine bite and sting injuries treated in U.S. emergency departments, 2001–2004. *Public Health Rep*. 2007;122:764–775.
9. Goddard J. Direct injury from arthropods. *Lab Med*. 1994;25:365–371.
10. Rolla G, Franco N, Giuseppe G, Marsico P, Riva G, Zanotta S. Cotton wool in pine trees. *Lancet*. 2003;361:44.
11. Wyatt JP, De Shazo R, Goddard J. Clinician's guide to common athropod bites and stings. In: Moraru GM, Goddard J, eds. *The Goddard Guide to Arthropods of Medical Importance*. 7th ed. Boca Raton, FL: CRC Press, Taylor and Francis Group; 2019:67–77.
12. Delgado A. Venoms of Lepidoptera. In: Bettini S, ed. *Arthropod Venoms*. Berlin: Springer-Verlag; 1978:555–611.

13. Diaz JH. The evolving global epidemiology, syndromic classification, management, and prevention of caterpillar envenoming. *Am J Trop Med Hyg.* 2005;72(3):347–357.

14. Mullen GR. Moths and butterflies (Lepidoptera). In: Mullen GR, Durden LA, eds. *Medical and Veterinary Entomology.* 2nd ed. New York: Elsevier; 2009:353–370.

15. Frazier CA. *Insect Allergy: Allergic reactions to Bites of Insects and Other Arthropods.* St. Louis: Warren H. Green; 1969.

16. Gamborgi GP, Metcalf EB, Barros EJ. Acute renal failure provoked by toxin from caterpillars of the species Lonomia obliqua. *Toxicon.* 2006;47(1):68–74.

17. CDC. Necrotic arachnidism—Pacific Northwest, 1988–1996. In: CDC, MMWR. 1996;45:433–436.

18. Diaz J. The global epidemiology, syndromic classification, management, and prevention of spider bites. *Am J Trop Med Hyg.* 2004;71:239–250.

19. Kemp ED. Bites and stings of the arthropod kind. *Postgrad Med.* 1998;103:88–94.

chapter twenty one

Fly Larvae in Humans (Myiasis)

Introduction and Medical Significance

When fly maggots infest tissues of people or animals the condition is referred to as myiasis.[1,2] Myiasis occurs only as a result of an egg-laying female fly (fly larvae are not capable of reproduction); therefore, myiasis is a stand-alone event and not contagious from patient to patient in a health-care setting. Specific cases of myiasis are clinically defined by the affected areas(s) involved, such as traumatic (wound), gastric, rectal, auricular, and urogenital myiasis, among others. Myiasis can be accidental, when fly larvae occasionally make their way into the human body, or facultative, when fly larvae enter living tissue opportunistically after feeding on decaying tissue in wounds. Myiasis can also be obligate, in which the fly larvae must spend a portion of their life cycle in living tissue. Obligate myiasis is the most serious form of the condition and constitutes true parasitism.

Accidental Myiasis

Accidental myiasis (also referred to as enteric myiasis or pseudomyiasis) is mostly a benign event clinically, but fly larvae could possibly survive long enough to cause stomach pains, nausea, or vomiting. Physicians and entomologists should be careful in confirming enteric myiasis, since many cases, some of which get into the scientific literature, are actually contamination of the toilet bowl or stool itself after the fact. Seeing maggots in the stool or toilet bowl is so alarming that patients may overlook other possibilities, such as soldier fly or drain fly larvae developing in the "scum" lining toilet pipes.

Nonetheless, some cases of accidental myiasis could be genuine.[3] I once consulted on a case wherein a 7-year-old child vomited up a fly larva 48 hours after eating plums and muscadines.[4] The specimen was retrieved by the mother and later identified by entomologists at Mississippi State University. Many different groups of flies may be involved in enteric myiasis, such as species in the families Muscidae, Calliphoridae, Sarcophagidae, Stratiomyidae, and Syrphidae. Notorious offenders are the cheese skipper, *Piophilia casei*; the black soldier fly, *Hermetia illucens* (Figure 21.1); and the rat-tailed maggot, *Eristalis tenax* (Figure 21.2). Other instances of accidental myiasis (nonenteric) occur when fly larvae enter the urinary passages or other body openings. Flies in the genera *Musca, Muscina, Fannia, Megaselia,* and *Sarcophaga* have been reported causing such cases.

Figure 21.1 Soldier fly larva, often reported in stool as a case of accidental myiasis.

Figure 21.2 Rat-tailed maggot.

TYPES OF MYIASIS (FLY LARVAE IN HUMAN TISSUES)

- Accidental (larvae accidentally ingested in food or drink)
- Facultative (opportunistic, wherein larvae infest malodorous or festering wounds)
- Obligate (true parasitism, wherein larvae *must* develop in living tissues)

Facultative Myiasis

Numerous species of Muscidae, Calliphoridae, and Sarcophagidae have been reported to cause facultative myiasis (Figures 21.3a and 21.3b), and may cause pain and tissue damage if fly larvae leave necrotic tissues and invade healthy tissues, even transiently. In the United States, the calliphorid *Lucilia sericata* has been reported causing facultative myiasis on several occasions.[5-7] During my career as a public health entomologist, I have identified *L. sericata* and other blow flies removed from living human tissues, such as the rectum, vagina, and eye socket (Figure 19.5). Another calliphorid, *Chrysomya rufifacies*, an introduced species, is also known to be regularly involved in facultative myiasis.[8] Other fly species commonly involved in this type of myiasis include: *Calliphora vicina*, *Phormia regina*, *Cochliomyia macellaria*, and *Sarcophaga haemorrhoidalis*.

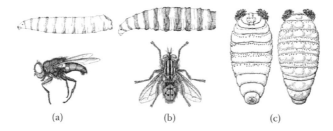

(a)   (b)   (c)

Figure 21.3 Some flies involved in myiasis: (A) Calliphoridae, (B) Sarcophagidae, and (C) Oestridae.

Source: Figure redrawn from several USDA sources

Figure 21.4 Human bot fly larva, third stage.

Source: Photograph copyright 2007 by Jerome Goddard, Ph.D.).

Obligate Myiasis

When certain fly species must develop in the living tissues of a host as part of their natural life cycle, the condition is called obligate myiasis. This type of fly infestation is mostly seen in wild and domestic animals, and veterinarians encounter it quite frequently. In humans, obligate myiasis usually occurs from screwworm flies (Old and New World) or human bot flies (Figure 21.3c and 21.4). Obligate myiasis from the human bot fly of Central and South America is rarely fatal, but myiasis from screwworm flies has often been reported to cause considerable pathology and human deaths. Screwworm flies use wild animals or livestock as primary hosts, but will seriously infest humans if given the chance. If, for example, a female screwworm fly oviposits just inside the nostril of a sleeping human, hundreds of developing larvae may migrate through the turbinal mucous membranes, sinuses, and other tissues. Surgical removal of all larvae in such cases is extremely difficult. Fortunately, screwworm flies have been eradicated from the United States.

More rarely, there are bot fly species that ordinarily infest wild animals, but which may only rarely attack humans. Such infestations may present clinically as a "maggot in a boil" or other furuncular-like lesion. Since the lesion develops in otherwise healthy tissue, and since there is often no international travel history, physicians may not be aware that other zoonotic forms of obligate fly parasites exist in the United States. In one such case that I investigated, a 3-year-old boy somehow became infested with a bot fly larva that normally attacks squirrels, chipmunks, or rabbits.[9] The family lived in a rural area near large tracts of woods containing abundant wildlife. According to the mother, the boy was stung by a bumble bee on his side and neck while watching television one morning. She said that within 5 minutes, typical sting-like "welts" occurred at the places the child pointed out. Within 2 days, a line of vesicles extended away from the lesions—presumably caused by the larvae migrating in the skin. The lesion on his side extended upward in a sinuous fashion about 10 cm, ending in a small papule, and developed no further (apparently the larva died). The larva in his neck continued to enlarge and migrated about 4 cm laterally. After 14 days, the lesion was an inflamed dermal tumor with a central opening about 3 mm in diameter, through which the larva obtained air. The child often cried and complained of severe pain. Despite numerous trips to physicians, the myiasis was not diagnosed until almost 4 weeks after the initial "stinging incident." An ER physician expressed the larva, which was ultimately forwarded to the state health department for identification. On examination, the specimen was identified as a second-stage larva of a fly in the genus *Cuterebra*, which is a member of the rabbit and rodent bot flies. Interestingly, adult *Cuterebra* flies look similar to bumble bees (Figure 21.5).

Figure 21.5 Adult *Cuterebra* bot fly.
Source: Photograph copyright 2014 by Jerome Goddard, Ph.D.

Contributing Factors

Accidental Myiasis

Fly eggs or young maggots on uncooked or previously cooked foods may be eaten by people, leading to accidental enteric myiasis. Infested foods often include cured meats, dried or unwashed fruits, cheese, and smoked fish. Other cases of accidental myiasis may result from using contaminated catheters, douching syringes, or other invasive medical equipment, or sleeping with the body exposed, particularly if the person has a urinary tract infection.

Facultative Myiasis

Many fly species oviposit on dead animals or rotting flesh, especially blow flies and flesh flies (Figure 21.6). Accordingly, these flies sometimes mistakenly lay eggs in a foul-smelling wound on a living animal or person, and these developing maggots can sometimes subsequently invade healthy tissue. Wounds with watery alkaline discharges (pH 7.1–7.5) seem especially attractive to blow flies. Facultative myiasis in humans frequently occurs in semi-invalids who have poor (if any) medical care. Often, in the case of the very elderly, their eyesight may be so weak that they do not detect the infestation. In clinical settings, facultative myiasis can occur in

incapacitated patients who have recently had major surgery or those having large or multiple uncovered or partially covered festering wounds.
I know of one case that involved fly larvae in a person's eye socket after
surgery (Figure 21.7). However, not all cases of facultative myiasis occur in
or near wounds. In the United States, larvae of the blow fly *L. sericata* have
been reported from the ears and nose of healthy patients with no other
signs of trauma in those areas.[10]

Figure 21.6 Dead pig with maggots. Blow flies are attracted to dead animals that
serve as sites for oviposition.

Figure 21.7 Fly larvae removed from a patient's eye socket several days post-surgery.

Source: Photograph copyright 2007 by Jerome Goddard, Ph.D.

Obligate Myiasis

In most cases of obligate myiasis, humans are not the ordinary hosts, but may become infested. Human infestation by the human bot fly is usually through a mosquito bite—the eggs are attached to mosquitoes and other biting flies; however, human screwworm fly myiasis may result from direct egg laying on a person, most often in or near a wound or natural orifice. I have known of several cases of screwworm infestation occurring in the human nose.

Myiasis in Clinical Practice

Physicians, other healthcare providers, and public health practitioners encounter cases of myiasis in a number of ways. Concerned patients may bring in larval specimens found in stool or in the toilet that they found and say "came out of them." Physicians and laboratory personnel should be careful not to confirm allegations of enteric myiasis without evidence. Just because someone found a maggot in their toilet bowl does not mean it passed out of their digestive tract. Depending upon several factors, including cleanliness of the home or bathroom, the maggots may have coincidentally been found in or near stool samples, or could subsequently have infested the stool samples. On a number of occasions, I have investigated cases of "maggots" in toilet bowls that were in fact soldier fly larvae (Figure 21.1) coming from the wax seal located where the toilet connects to the floor. When the larvae of this particular fly species are ready to pupate, they crawl away from their food source and often end up in the toilet bowl.

Most clinical myiasis samples are from facultative myiasis originating from blow fly (Diptera: family Calliphoridae) larvae being found in a patient's nose, ear, rectum, or pus-filled wound. If the site is a natural orifice, there is or was usually a lesion or infection that proved attractive to the female fly. Many times, the patient is an invalid and unable to care for himself or herself. These cases of myiasis are usually not life threatening because the larvae rarely invade healthy tissue. Identification of the larvae in clinical samples is fairly easy to the family level (blow fly, house fly, flesh fly, etc.) using algorithms, keys, and illustrations available on the Internet. For example, house fly larvae have light brown posterior spiracles located on flat or rounded areas *without* any ring of soft protuberances around them (Figure 21.8). Blow fly and flesh fly larvae, on the other hand, have their spiracles set in depressions surrounded by a ring of soft fleshy protuberances. Patients with myiasis should be questioned about their recent travel history. Occasionally, human bot fly or screwworm myiasis occurs in travelers returning from tropical countries. One woman I knew complained about a "boil" occurring behind her ear after

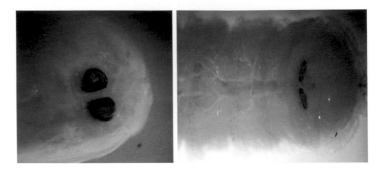

Figure 21.8 Posterior spiracles of the house fly (left) versus blow fly larva (right).

a trip to Belize. She claimed she could feel or sense a "clicking" sound inside the boil. Eventually she was seen by a physician who diagnosed human bot fly myiasis (Figure 21.4). Apparently, she really was sensing the larva as it moved or fed in the tissues near her ear.

Differential Diagnosis

Cases of facultative myiasis are fairly obvious—maggots clearly visible in wounds or sores—but obligate myiasis from bot flies may be more difficult to diagnose. Boil-like or nodular lesions on human skin can result from many things, including staphylococcal infections, cat-scratch disease, tick-bite granuloma, tungiasis (a burrowing flea), and infestation with various parasitic worms (such as *Dirofilaria*, *Loa loa*, and *Onchocerca*), as well as many other causes. Nodular lesions eventually ulcerate if the inflammatory process is intense enough to destroy the overlying epidermis, but lesions from myiasis do not ulcerate. The central core of the lesion should be examined for evidence of a fly larva, as the larva sometimes can be clearly seen just below the skin surface. Another helpful clue in diagnosing myiasis with the human bot fly, *Dermatobia hominis*, is that sometimes the pointed posterior end of the larva protrudes from the central opening as high as 5 mm above the skin.

Prevention, Treatment, and Control

Accidental and facultative myiasis can often be averted by prevention, sanitation, and personal protection measures. Food should not be exposed or unattended for any length of time to prevent flies from ovipositing therein. Covering and refrigerating leftovers should be done immediately after meals. Washing fruits and vegetables prior to consumption can help remove developing maggots, although a visual examination should also be performed when slicing or preparing those items. Other forms of accidental myiasis may be prevented by covering and protecting invasive medical equipment from flies and avoiding sleeping nude, especially during daytime. For facultative myiasis, extra care should be taken to keep wounds clean and covered, especially in elderly or invalid individuals. Regular visits by a home health nurse can help prevent facultative myiasis in patients who stay at home. For those institutions containing vulnerable patients, every effort should be made to prevent entry of flies into the facility. This might involve exclusion methods such as keeping doors and windows screened and in good repair, thoroughly sealing all cracks and crevices, installing air curtains over doors used for loading and unloading supplies, and installing UV fly traps in areas accessible to the flies, but inaccessible to patients. Prevention of obligate myiasis involves avoiding sleeping outdoors during daytime in screwworm-infested areas (tropics) and using insect repellents in Central and South America to prevent bites by bot fly egg-bearing mosquitoes.

Treatment of accidental enteric myiasis is probably not necessary since in most cases there is no development of fly larvae within the highly acidic stomach environment and other parts of the digestive tract. They are usually killed and passively carried along through the digestive tract. Treatment of other forms of accidental myiasis as well as facultative or obligate myiasis involves direct removal of the larvae. Alexander[11] recommends debridement with irrigation. Others have suggested surgical exploration and removal of fly larvae under local anesthesia.[10] Fly larvae should be handled carefully during extraction procedures to avoid bursting them (this could cause an allergic reaction). "Bacon therapy" has been used successfully to remove human bot fly larvae. This method involves covering the punctum (breathing hole in the patient's skin) with raw meat or pork.[12] In a few hours, the larvae migrate into the meat and are then easily extracted when the meat is removed. Maggot infestation of the nose, eyes, ears, and other areas may require surgery if larvae cannot be removed directly from natural orifices. Since blow flies and other myiasis causing flies lay eggs in batches, there could be tens or even hundreds of maggots in a wound, making their removal problematic.

References

1. James MT. The flies that cause myiasis in man. In: U.S. Dept. Agri. Misc. Publ. No. 631, Washington, DC; 1947:175 pp.
2. Scholl PJ, Catts EP, Mullen GR. Myiasis (Muscoidea, Oestroidea). In: Mullen GR, Durden LA, eds. *Medical and Veterinary Entomology.* 2nd ed. New York: Elsevier; 2009:309–338.
3. Mazzotti L. Casos humanos de miasis intestinal. *Ciencia, Mex.* 1967;5:167–168.
4. Goddard J, Hoppens K, Lynn K. Case report on enteric myiasis in Mississippi. *J Mississippi St Med Assoc.* 2014;55(4):132–133.

5. Greenberg B. Two cases of human myiasis caused by *Phaenicia sericata* in Chicago area hospitals. *J Med Entomol.* 1984;21:615.

6. Merritt RW. A severe case of human cutaneous myiasis caused by *Phaenicia sericata. California Vect Views.* 1969;16:24–26.

7. USDA. Pests infesting food products. In: United States Department of Agriculture, ARS, Agri. Hndbk. No. 655; 1991:213 pp.

8. Richard RD, Ahrens EH. New distribution record for the recently introduced blow fly *Chrysomya rufifaces* in North America. *Southwest Entomol.* 1983;8:216–218.

9. Goddard J. Human infestation with rodent botfly larvae: a new route of entry? *South Med J.* 1997;90:254–255.

10. Anderson JF, Magnarelli LA. Hospital acquired myiasis. *Asepsis.* 1984;6:15.

11. Alexander JO. *Arthropods and Human Skin.* Berlin: Springer-Verlag; 1984.

12. Brewer TF, Wilson ME, Gonzalez E, Felsenstein D. Bacon therapy and furuncular myiasis. *J Am Med Assoc.* 1993;270:2087–2088.

Index